Understanding
the Mathematics of
Personal Finance

Understanding the Mathematics of Personal Finance

An Introduction to Financial Literacy

Lawrence N. Dworsky

A John Wiley & Sons, Inc., Publication

Published by John Wiley & Sons, Inc., Hoboken, New Jersey.
Published simultaneously in Canada.

For general information on our other products and services or for technical support, please contact our
Customer Care Department within the United States at (800) 762-2974, outside the United States at
(317) 572-3993 or fax (317) 572-4002.

Wiley also publishes its books in a variety of electronic formats. Some content that appears in print
may not be available in electronic formats. For more information about Wiley products, visit our web
site at www.wiley.com.

Library of Congress Cataloging-in-Publication Data:

Dworsky, Lawrence N., 1943–
    Understanding the mathematics of personal finance / Lawrence N. Dworsky.
        p. cm.
    Includes bibliographical references and index.
    ISBN 978-0-470-49780-7 (pbk.)
    1. Finance, Personal–Mathematics. 2. Investments–Mathematics. 3. Business mathematics.
4. Consumers–Decision making–Mathematics. I. Title.
    HG179.D92 2009
    332.024001'5195–dc22

                                                                                    2009015925

10 9 8 7 6 5 4 3 2 1

To all the people struggling to understand the calculations behind the various financial instruments they encounter: I hope this book helps.

Contents

Preface

What is personal finance? An informal definition is "how you interact with money." Among the subcategories of personal finance are topics such as budgeting, saving, borrowing, investing, gambling, and buying and selling real estate. Many books, courses, professional advisors, and software programs are available to help you optimize your path through your financial life.

This book is about various forms of borrowing and saving money, and includes some discussion of investing money. Borrowing money takes many forms, including home mortgage loans, auto loans, and credit card debt. Saving money includes putting money under your mattress, depositing it into a savings bank, and buying certificates of deposit (CDs). Insurance policies can be thought of as a special kind of pooled savings plan whereby many people put money into the same savings account, and this money becomes available to these people when a specified special need (illness, repairing a car, death benefit) unexpectedly arises. Investing is an opportunity to earn more money with your money than a savings bank will give you, but with less certainty about the earnings and, for that matter, less certainty about maintaining your original money than a government-insured savings account would give you.

When you borrow money or, equivalently, take a loan from a person, a bank, a mortgage company, or elsewhere, you will be expected to pay a fee for the use of this money. The amount you borrow is called the *principal* of your loan and the fee you pay for borrowing the money is called the *interest*. The amount of interest you have to pay is based upon the principal, the amount of time you have the money, and the prevailing financial conditions. The longer you have this money, the more interest you can expect to pay. In common situations such as a home mortgage or a car loan, you usually repay the loan gradually over a period of time. In this case, calculating the interest gets a little messy because the amount you owe at any given time (the *balance*) is being reduced due to your payments, while it is simultaneously being increased by the accrual of interest based on your balance at that time. In a properly structured loan, your payments are large enough that the balance decreases after each payment and eventually goes to 0, so that your loan is *paid off*.

The concepts and calculations for a simple one repayment loan and for multiple payment loans such as mortgages and car loans are the same; it's just that in the latter cases you have to repeat the same calculations many times. Before the era of spreadsheets on personal computers and the Internet, the complexity of the multiple calculations was so significant that only banks and mortgage companies and other large financial institutions could undertake them. When you took a loan, you would be provided with a table of payment due dates and loan balances (an *amortization* table) for your loan. Comparing different loan opportunities was very difficult unless you wanted to spend a lot of time in the library working with books of loan tables.

Today, everybody can easily calculate loan details themselves. Pocket calculators with all the necessary financial functions built-in are inexpensive and easy to use. Users of spreadsheet programs on personal computers can generate their own amortization tables based on the financial functions built into these spreadsheets and/ or can build up these formulas from basic principles. Most common financial calculations are available on the Internet ("online") in the form of simple calculators designed specifically for a single type of problem.

My goal in writing the book is to explain how even the most involved loan scenarios can be understood just by repeated application of the fundamental concept of *compound interest*, which is the subject of Chapter 2. I'll show how to calculate everything involved with these loans using a computer spreadsheet program, and whenever possible, I'll reference some online calculators—particularly those on my own website.

I should mention here that I'm using "loan" as a generic term for one party letting another party use his or her money for some time and expecting interest as compensation. When you take a mortgage loan on a home, you are borrowing the money from somebody. When you put money into a savings bank or purchase a CD ("invest" your money), the bank is borrowing money from you. In terms of the mathematics involved, these are identical situations—you just have to keep track of which way dollars are flowing.

If you *loan* me money, then I am *borrowing* money from you and vice versa. In terms of usage, I often see that the terms *loan* and *borrow* are used interchangeably. In many situations that you encounter, you'll simply have to pull the correct meaning out of context. This is unfortunate because each term has a specific meaning; they're not interchangeable. I will admit that the correct usage can sometimes be confusing—when I *take a loan*, I'm *borrowing* money. The person or company that loaned me the money is the *lender*, and once I've borrowed the money I am the *debtor*.

This is not a book that gives investment or borrowing strategies. I won't offer suggestions on how to plan for retirement, whether or not you want a reverse mortgage, how to allocate your savings, and so on. My goal is to provide the tools for you to be able to calculate the real costs and/or profits involved in using these various financial instruments and therefore to put you in a position to see for yourself what the best deals are and/or how you could sometimes get yourself into a financial mess.

The most important concept to hold in your mind is that because of interest accruing on borrowed money, the amount of money you owe (or are owed) has a *time value* to it. One thousand dollars to be paid to you today is worth more than $1,000 to be paid to you a year from now. One thousand dollars to be paid to you a year from now is worth more than $2,000 to be paid to you 20 years from now. You must learn to work with concepts such as *future value*, which is the amount that some number of dollars today will be worth on a specific date in the future, and also the *present value*, which is the amount that some number of dollars on some specific date in the future is worth today.

In this book, you will find descriptions of various financial instruments (mortgages, credit card purchases, cash advances, etc.) You will also see how these financial instruments work and how to use the proper analysis tools (primarily the computer spreadsheet) well enough that you can tackle a new situation and come up with the right answers.

There are many computer spreadsheet programs available. Fortunately, they are all very similar in structure, and the instructions I give for my spreadsheets will work on all popular spreadsheet programs.

The spreadsheet calculators used in this book are all available on my website (www.lawrencedworsky.com). Chapter 15 shows you how to get a free spreadsheet program if you need one, how to get to the spreadsheets I'm providing, and a general introduction on how I'm setting them up and how to use and maintain them.

In a sense, this book will never be finished. My website will always be changing. I will improve the existing spreadsheets, adding examples and explanations as well as new capabilities. I will have an up-to-date *errata* section (that hopefully will be very, very, short). Also, my website has the typical Contact Me capability. This is how I will learn what I haven't explained well, what relevant facts or scenarios I have overlooked, and so on. I will address all of these matters and put my work on the website as quickly as possible. Interesting problems may become additional problems for the book, posted on the website.

Chapter 1 contains a review of the basic mathematics necessary to understand the book. Most readers shouldn't find this math difficult. The only new information presented is that the notation isn't usually what was taught to you in high school. I'll go through this slowly and carefully. There are powerful notations to properly express calculations that you probably already know how to do. These notations are important because they can describe involved calculations clearly and concisely.

Included in the book are a few sections of mathematical nature that delve a bit more deeply into a topic than does most of the book. These sections are not necessary for a good understanding of the book or use of the calculator spreadsheets and can be skipped if you wish. I'll clearly state at the beginning of each of these sections that you can skip the section if you don't want to wrestle with the mathematics.

LAWRENCE N. DWORSKY

Acknowledgments

A tolerant group of relatives and friends helped me to interpret various published documents about different financial instrument rules' calculations and then read my drafts and commented on whether or not I was explaining things more clearly. This group includes my wife Suzanna, my daughter and son-in-law Gillian and Aaron Madsen, and my friends Mel Slater and Chip Shanley.

Susanne Steitz-Filler at John Wiley and Sons has been patient and helpful as the structure of this book evolved from my original ideas.

I thank you all.

List of Abbreviations

ADB	Average Daily Balance
APR	Annual Percentage Rate
ARM	Adjustable Rate Mortgage
ATM	Automatic Teller Machine
CD	Certificate of Deposit
Cmpds	Compoundings
CORR	Corrected
COL	Cost of Living
DB	Daily Balance
EAPR	Effective Annual Percentage Rate
HECM	Home Energy Conversion Mortgage
HUD	(Department of) Housing and Urban Development
IAWPC	Immediate Annuity with Period Certain
INC	Income
Int or INT	Interest
IOU	I Owe You (a promissory note)
IRA	Individual Retirement Account
Mnth	Month
NAPR	Nominal Annual Percentage Rate—same as APR
Nr	Number
NrPmts	Number of Payments
PMT	Payment
PV	Present Value
SEC	(U.S.) Securities and Exchange Commission
SEP	Single Employer Pension
Tot or TOT	Total
Vol	Volume (number of shares traded in a given time)

Chapter 1

Background Mathematics

1.1 ARITHMETIC, NOTATION, AND FORMULAS

Almost all of the mathematics used in this book involve only the four basic operations of addition, subtraction, multiplication, and division. If you can comfortably read about and then actually perform calculations using these four operations, you have all the math background you need. If you have a pocket calculator or a computer with a spreadsheet program, then you have the "machine power" to do whatever you need to do without resorting to pencil and paper.

Mathematical *notation*, the way we express what we want to calculate, can sometimes be confusing. Mathematical notation is the vocabulary of the language of mathematical concepts. Often a student will think he or she doesn't understand a concept when he or she simply is not familiar with the notation. It's like being given driving directions in a foreign language when you don't know the words for "turn left" or "turn right." To further complicate things, there is almost always more than one way to write a particular mathematical expression. Often the choice is a matter of style and/or convenience. In this section, I'll go through the various ways of writing different expressions involving only the four basic operations and explain why I will choose what I choose when I choose it.

I'll begin with the definition of a *variable*. A variable, simply speaking, is a letter or a name that represents a number. If I want to say, for example, that an item in a catalog costs the price listed in the catalog plus a $10 shipping and handling fee, I can write, "If the cost of an item is X dollars, then you must pay $X + 10$ dollars if you want to order any item from this catalog." It's a way of generalizing a relationship rather than having to recreate the relationship for each example.

I can get fancier and say that if Y represents the amount you must pay, then

$$Y = X + 10.$$

Some people like to use letters from the end of the alphabet; some like to use letters from the beginning of the alphabet; some like to use Greek letters. It doesn't matter which letters are used, as long as you're clearly told what number each of

Understanding the Mathematics of Personal Finance: An Introduction to Financial Literacy, by Lawrence N. Dworsky
Copyright © 2009 John Wiley & Sons, Inc.

these letters represent. Some authors of books and computer programs use variables that are case sensitive. That is, X and x represent different numbers. I don't do this either in this book or in my spreadsheets.

In some situations, it's convenient to use a whole word as the variable. In the above example, instead of letting Y be the cost, I could have written

$$Cost = X + 10.$$

The expression

$$Y = X + 10$$

is called a *formula*. A formula is a mathematician's version of a recipe. You put in X (in this case the catalog price for the item) and you get out Y (in this case the amount you must pay to have the item appear on your doorstep). It's conventional, but not necessary, for the variable that you're calculating to appear on the left-hand side of the equal sign and the variable(s) that you're supplying to appear on the right-hand side of the equal sign.

Typically, numbers that don't change are shown as numbers, such as the 10 in the above formula, and numbers that depend on your particular situation, such as X and Y in the above formula, are represented by letters. This isn't a law; it's just a common practice.

If, for example, the shipping cost depends on the item's weight, I could say that

$$Y = X + Z,$$

where Z is the shipping cost. If I do this, then I must refer you to a table or to another formula that explains how to calculate or look up the shipping cost before you can use this formula.

I'll start my discussion of the four basic operations with addition, spelling out some things that are probably obvious. I'm doing this in order to be able to draw a contrast with the other basic functions and also to start explaining the use of parentheses.

When adding numbers, it doesn't matter what order you do things in:

$$3 + 7 + 5 = 3 + 5 + 7 = 7 + 5 + 3 = \ldots$$

Note also the use of the expression "...". This means "and so on" and hopefully will be obvious in its intent when I use it.

I could also write the above addition example using parentheses to group some of the operations:

$$(3 + 7) + 5 = .$$

Writing it this way means "add 3 to 7 first, then add the result to 5." In this example, the parentheses don't contribute any value since the order of the additions doesn't matter. On the other hand, they don't introduce any error. In short, in this example, while the parentheses are harmless, they're also pointless.

Now, let's look at subtraction:

$$7 - 3 = 4.$$

This is pretty clear so far.

However, while 2 + 6 is the same as 6 + 2,

$$6 - 2 = 4 \text{ and } 2 - 6 = -4$$

are clearly not the same.

Getting a little more complicated,

$$(7 - 3) + 2 = 6.$$

The parentheses here mean first evaluate 7 − 3 (= 4) and then add the result to 2, yielding 6.

This is not the same as

$$7 - (3 + 2) = 2.$$

In this case, the instructions are to first add 3 + 2 (= 5) and then subtract the result from 7, yielding 2.

Subtraction differs from addition in the importance of notation because the order in which things are calculated matters.

If I were to just write

$$7 - 3 + 2 = ??,$$

I wouldn't know how to evaluate this because without the instructions added by the parentheses, I just don't know what to do first.[1]

Moving on to multiplication, the simplest notation (and one that's hardly ever used) is to use an "×" to indicate multiplication. Using this notation, it's clear to see that, as in the case of addition, order doesn't matter:

$$3 \times 2 \times 6 = 3 \times 6 \times 2 = 6 \times 2 \times 3 = \ldots = 36$$

One good reason why the "×" is hardly used to signify multiplication is that, once you're expecting formulas, you don't know whether this × means multiplication or is itself a variable representing another number.

For better or worse, there are many notations for multiplication. The important consideration is that the chosen notation must be clear and unambiguous.

When it is clear what I mean, I will just write the two numbers (and/or variables) that I want to multiply next to each other: $3x$, or xy. Obviously this won't work for multiplying 3 by 2, because 32 (or 23) would be interpreted as a two-digit number, not instructions to multiply the two single-digit numbers together.

When multiplying a number by a variable, it's common to put the number first: $3x$ means the same as $x3$ but is almost always written as $3x$.

This example also shows why multiplication is almost never written using an "x" to signify multiplication—the "x" is probably the most common choice of a letter for a variable, and writing $3 \times x$ to mean "multiply 3 by the variable x" is just a confusing mess.

---

[1] Computer programming languages usually resolve this kind of ambiguity by having a default procedure such as "when there are no explicit instructions (parentheses), work from left to right." I won't assume any such default procedures in this book.

The expression

$$3x(y+7)$$

can be interpreted two ways. The two ways are equivalent and both are valid.

The first interpretation is that you should do what's inside the parentheses first. That is, if y represents some cost or payment or whatever, let's say $y = \$15.50$, then add y to 7, giving $15.50 + 7 = 22.50$. Then we have

$$3x(22.50).$$

This is simply three numbers multiplied together. The parentheses now are used just to keep the 3 from being tangled up with the 22.50. Since numbers multiplied together can be multiplied in any order, we have

$$3x(22.50) = 3(22.50)x = 67.50x$$

At this point we need a value for x or we just have to stop.

In order to do what I just did, I needed a value for the variable y. If I don't have a value for y, I can either leave things as they are for the time being, or I can "expand" the expression. This is the second interpretation: What's outside the parentheses multiplies everything that's inside the parentheses. Therefore,

$$3x(y+7) = 3x(y) + 3x(7) = 3xy + 21x$$

Whether the latter way of writing things is any clearer, or more useful, than the original expression is in the eye of the beholder.

Taking this one step further, what if I have

$$(12+4)(3+6).$$

The same rules apply; you just have to do a little more work:

$$(12+4)(3+6) = (16)(9) = 144.$$

This type of expression, when there are algebraic variables involved, often trips up students. Another correct way of evaluating this expression is to use the second interpretation above: What is outside the parentheses multiplies everything that is inside the parentheses. This time we have to remember that there are two sets of parentheses, so we have

$$(12+4)(3+6) = 12(3+6) + 4(3+6) = 12(3) + 12(6) + 4(3) + 4(6)$$
$$= 36 + 72 + 12 + 24 = 144.$$

This is sometimes called "expanding" the expression.

An example of the same expression with algebraic variables is

$$(a+b)(c+d) = ac + ad + bc + bd.$$

In this book, I'll often be presenting formulas for use in calculating a number, typically a dollar value. This isn't an algebra book. You will have the working knowledge that you need as long as you understand the first interpretation, that is, put in the values for the variables then evaluate what's inside the parentheses (or sets of parentheses).

The last of the four basic operations is division. First, notation: The elementary school notation $6 \div 3$ is pretty much never used. Instead, recognizing that a division expression is the same as a fraction expression, 6 divided by 3 will be written as one of the following:

$$6/3 = \frac{6}{3} = 2.$$

If I want to use the first of these forms for multiple operations, then I have to get involved with parentheses, because order counts. That is,

$$18/6/3 =$$

is ambiguous because I don't know what to do first. My choices are

$$(18/6)/3 = 3/3 = 1$$

or

$$18/(6/3) = 18/2 = 9,$$

and I have no way of knowing which interpretation was intended.

Using the fraction notation, I can finesse the parentheses issue by working with different size fraction lines. That is,

$$\frac{\frac{18}{6}}{3} = \frac{3}{3} = 1$$

while

$$\frac{18}{\frac{6}{3}} = \frac{18}{2} = 9.$$

I could keep going with compound expressions—say, fractions involving sums or differences of numbers and variables inside the parentheses, and so on. But again, this is not an algebra book and my aim is not to trip you up but to give you clear rules for evaluating (finding the numeric value of) a formula when you're presented with it.

1.2 MINUS (NEGATIVE) SIGNS

The seemingly benign set of rules for manipulating a minus sign nevertheless manages to cause an endless set of headaches. Let's see if I can summarize these rules quickly and clearly:

1. (Not so much a rule as a reminder) When a sign is not shown, a positive sign is implied:

$$34 = +34,$$
$$(35) = (+35) = +(35).$$

2. Subtracting B from A is the same as adding −B to A:

$$7 - 5 = 7 + (-5) = 2.$$

3. As implied above, multiplying a positive number by a negative number yields a negative number:

$$(-3)(6) = -(3)(6) = -18,$$
$$(-6) = -(6) = (-1)(6).$$

4. Multiplying two negative numbers yields a positive number:

$$(-5)(-7) = +35.$$

5. Division rules are the same as multiplication rules. Dividing a positive number by a negative number or dividing a negative number by a positive number yields a negative number. Dividing a negative number by a negative number yields a positive number:

$$\frac{4}{-3} = \frac{-4}{3} = -\left(\frac{4}{3}\right) = -\frac{4}{3} = -1.33,$$

$$\frac{-4}{-3} = \frac{4}{3} = 1.33.$$

1.3 LISTS AND SUBSCRIPTED VARIABLES

Throughout this book, I make frequent use of tables. Tables are lists of numbers that relate variables in different situations. This isn't as bad as it first sounds. I'm sure you've all seen this many times—everything from income tax tables that the Internal Revenue Service provides to automobile value depreciation tables.

Table 1.1 is a hypothetical automobile value depreciation table. Don't worry about what kind of car it is—I just made up the numbers for the sake of this example.

Looking from left to right, you see two columns: the age of the car and the car's wholesale price. Looking from top to bottom you see six rows. The top row contains the headings, or descriptions, of what the numbers beneath mean. Then there are

Table 1.1   Hypothetical Automobile
Value Depreciation Table

Age of car (years)	Wholesale price ($)
0	32,000
1	26,500
2	21,300
3	18,000
4	15,500
5	13,250

Table 1.2 Hypothetical Automobile Depreciation Table
with Air-Conditioning Option

Age of car (years)	Wholesale price ($)	Extra for air-conditioning
0	32,000	1,200
1	26,500	1,050
2	21,300	850
3	18,000	650
4	15,500	550
5	13,250	450

five rows of numbers. The numbers on each row "belong together." For example, when the car is 2 years old, the wholesale price is $21,300.

An important point about the headings is that whenever appropriate, the *units* should be listed. In Table 1.1, the age of the car is expressed in years. If I didn't say so, how would you know I didn't mean months, or decades? The value of the car is expressed in dollars. To be very precise, maybe I should have said U.S. dollars (if that's what I meant). Someone in Great Britain could easily assume that the prices are in pounds if I didn't clearly state otherwise.

Very often a table will have many columns. Table 1.2 is a repeat of Table 1.1, but with a third column added: How much more the car is worth if it has air-conditioning. Notice that I was a little sloppy here. I didn't say that the extra amount was in dollars. In this case, however, a little sloppiness is harmless. Once you know that we're dealing in dollars, you can be pretty sure that things will be consistent.

Again, the numbers in a given row belong together: A 3-year-old car is worth $18,000, and it is worth $650 more if it has air-conditioning.

Tables 1.1 and 1.2 tell you some dollar amounts based on the age of the car. It's therefore typical for the age of the car to appear in the leftmost column. I could have put the car's age in the middle column (of Table 1.2) or in the right column. Even though doing this wouldn't introduce any real errors, it makes things harder to read.

Whenever convenient, columns are organized from left to right in order of decreasing importance. That is, I could have made the air-conditioning increment the second column and the car value the third column (always count columns from the left), but again it's clearer if I put the more important number to the left of the less important number.

Some tables have many, many rows. The Life Tables presented in Chapter 10, the chapter about life insurance, have 102 rows—representing ages from 0 to 100, plus the heading row. The second column in the Life Tables is a number represented by the variable q, the third by the variable l, and so on. Don't worry about what these letters mean now; this is a topic in Chapter 10.

In Table 1.3, I've extracted a piece of the Life Table shown in Table 10.1. As you can see, for every age there are six associated pieces of information. Suppose I wanted to compare the values of q for two different ages, or to make some

Table 1.3   Part of the 2004 U.S. Life Table for All Men

Age	q	l	d	L	T	e
0	0.007475	100,000	747	99,344	7,517,501	75.2
1	0.000508	99,253	50	99,227	7,418,157	74.7
2	0.000326	99,202	32	99,186	7,318,929	73.8
3	0.000250	99,170	25	99,157	7,219,744	72.8
4	0.000208	99,145	21	99,135	7,120,586	71.8
5	0.000191	99,124	19	99,115	7,021,451	70.8
6	0.000182	99,105	18	99,096	6,922,336	69.8
7	0.000171	99,087	17	99,079	6,823,240	68.9
8	0.000152	99,070	15	99,063	6,724,161	67.9
9	0.000125	99,055	12	99,049	6,625,098	66.9
10	0.000105	99,043	10	99,038	6,526,049	65.9
11	0.000111	99,033	11	99,027	6,427,011	64.9
12	0.000162	99,022	16	99,014	6,327,984	63.9
13	0.000274	99,006	27	98,992	6,228,970	62.9
14	0.000431	98,978	43	98,957	6,129,978	61.9
15	0.000608	98,936	60	98,906	6,031,021	61.0
16	0.000777	98,876	77	98,837	5,932,116	60.0
17	0.000935	98,799	92	98,753	5,833,278	59.0
18	0.001064	98,706	105	98,654	5,734,526	58.1
19	0.001166	98,601	115	98,544	5,635,872	57.2
20	0.001266	98,486	125	98,424	5,537,328	56.2
21	0.001360	98,362	134	98,295	5,438,904	55.3
22	0.001419	98,228	139	98,158	5,340,609	54.4

generalizations of some sort. As I go through my discussion, I find that it's very cumbersome repeating terms like "the value of q for age 10" over and over again.

I can develop a much more concise, easy to read, notation by taking advantage of the fact that the left-hand column is a list of nonrepeating numbers that increase monotonically. By this I mean that 1 is below 0, 2 is below 1, 3 is below 2, and so on, so that it's easy to understand what row I'm looking at just by referring to the age (the left-hand column). Then I use a subscript (a little number placed low down on the right) tied to any variable that I want to discuss to tell you what I'm looking at. This is hard to describe but easy to show with examples:

q_3 refers to the value of q for age 3: $q_3 = 0.000250$.

q_{12} refers to the value of q for age 12: $q_{12} = 0.000162$.

d_{15} refers to the value of d for age 15: $d_{15} = 60$.

Now I can easily discuss the table using this subscript notation. In Table 1.3, q_{10} is the smallest of all the values of q, l_{22} is about 2% smaller than l_0, and so on. Asking why I'd want to be saying these things depends on the topic and the table under discussion. It's like asking why I'd ever want to multiply two numbers together.

1.4 CHANGES

When a number that you're interested in (the cost of a pound of coffee or the cost of a new home) changes, it's often more relevant to look at the percent change than it is to look at the absolute numbers.

For example, if you've been paying $3.00 a pound for coffee and the price changes by $2.00 up to $5.00 a pound, this is a relatively big change. On the other hand, if you've been considering purchasing a new car for $25,000 and the price changes by $2.00 to $25,002, relatively speaking, this is not a big difference.[2]

The standard way of calculating percent change is by subtracting the new value from the old value, and then by dividing this difference by the old value:

$$\% \text{ Change} = 100 \frac{\text{New value} - \text{Old value}}{\text{Old value}}.$$

The 100 multiplier is just to change the fractional quantity into a percentage.

Example: The pound of coffee mentioned above. The price was (the old value) $3.00, and the price now is (the new value) $5.00 so that

$$\% \text{ Change} = 100 \frac{\$5.00 - \$3.00}{\$3.00} = 100 \left(\frac{2}{3} \right) = 66.7\%.$$

Remember that the units of the new and old values must be the same. Don't have the new value in pennies, or francs, or any other units, if the old value is in dollars (and vice versa). Also, because we're dividing dollars by dollars, or francs by francs, the percent change is called *dimensionless*. It has no units. This is reassuring, because if we were to first convert our dollar amounts into, say euros, and then calculate the percent change, we should certainly expect to get the same answer.

If the new value is smaller than the old value, then the percent change will be a negative number.

Example: The price of a new car dropped from $30,000 to $27,000. The percent change is

$$\% \text{ Change} = 100 \frac{\$27,000 - \$30,000}{\$30,000} = 100 \frac{-3}{10} = -30\%$$

Looking at changes as percent changes or fractional changes (just don't multiply by 100) helps us to put things in perspective. We are comparing how much a number changes to how much the number used to be.

Percent change is not symmetrical in that changing a number and then changing it back doesn't give you the same results. For example, if I have something that used

---

[2] You might ask why not be concerned about the car price change—it's the same $2. As I see it, the difference is that you'll have your new car about 5 or 8 years, but you buy a pound of coffee every few weeks.

to cost $100 and now it costs $150, the percent change was 50%. However, if the price then returns from $150 back to $100, we've reversed the titles of old and new values. In this latter case, the percent change is −33.3%.

In some situations we don't have an old value or a new value. Consider the statement "My bank account seems to swing back and forth between $700 and $800." How do you calculate the percent change?

In this case, where the numbers seem to be taking turns being the old and new numbers, it makes sense to calculate the average of the two numbers,

$$\text{Average} = \frac{\$700 + \$900}{2} = \$800,$$

and then to talk about a percent variation or sometimes a "percent swing" by subtracting the smaller number from the larger number and then dividing the result by this average:

$$\% \text{ Variation} = 100 \frac{\$900 - \$700}{\text{Average}} = 25\%.$$

In this case, the two percent change numbers about the average are

$$100 \frac{\$900 - \$800}{\$800} = 12.5\%$$

and

$$100 \frac{\$700 - \$800}{\$800} = -12.5\%.$$

This is often written as ±12.5%, which is read as "plus or minus 12.5%."

One last little item that makes writing and reading about small changes more convenient is that a small change in a number (not a percent change) is often denoted by the Greek uppercase letter delta (Δ). If, for example, we have a cost of something that changes from an old value of $125.00 to a new value of $127.00, we would write

$$\Delta \text{Cost} = \$127.00 - \$125.00 = \$2.00.$$

As above, the convention is to subtract the old value from the new value. If the new value is smaller than the old value, the result is negative. Note that ΔCost has the same units as the two numbers used to calculate it (in this example, dollars), and consequently, the two numbers must have the same units (both be dollars, or both be cents, or francs, or euros, and so on).

Sometimes, it's necessary to work these problems backward. Suppose I tell you that a store is having a "30% off sale" on all items. What would the sale price be on an $80 purse? What we're looking for here is

$$100 \frac{\text{New price} - \$80}{\$80} = -30\%.$$

I'm sure I could awe you with my prowess at algebraic manipulation, but it's really not called for. This can easily be solved in either of two ways:

1. If we're reducing the price by 30%, then we're keeping 70% of the original price: 0.7($80) = $56.

2. Thirty percent of $80 is $24. Reducing an $80 price by $24 leaves $80 − $24 = $56.

These simple calculations work correctly because we're looking for the new price. If we have to go the other way, the problem is a bit trickier. For example, "A purse that was reduced by 30% is now selling for $56. What was the original price?"

Without trying to awe you (or more likely to bore you) with the derivation, the formula you need here is

$$\text{Old price} = \frac{\text{New price}}{\frac{\% \text{ Change}}{100} + 1}.$$

Putting in the appropriate numbers,

$$\text{Old price} = \frac{\$56}{\frac{-30}{100} + 1} = \frac{\$56}{-0.3 + 1} = \frac{\$56}{0.7} = \$80.$$

1.5 EXPONENTS

An exponent is another neat notation. Suppose I want to multiply an expression by itself (called "squaring" the expression):

$$(j+7)(j+7) = (j+7)^2.$$

The little "2" placed high up in the upper right means "square the expression" or more directly, "write the expression down twice, making it clear that you mean multiplication." This is also sometimes called "raising the expression to the power of 2."

Similarly, I can "cube" the expression

$$(j+7)(j+7)(j+7) = (j+7)^3$$

and so on.

In general, (anything)^n is called raising the expression "anything" to the nth power.

The following discussion of exponents is not needed in order to understand the book; I just thought that some readers might be curious as to why raising an expression to the power of 2 is called "squaring the expression" and raising it to the power of 3 is called "cubing the expression."

The area inside a rectangle is calculated by multiplying the rectangle's length by its width. A square is a rectangle whose length is equal to its width. In other words, all four sides of a square are the same length. The area of a square is therefore calculated by taking the square's length (or width) and multiplying it by itself. Consequently multiplying a number by itself is called "squaring" and raising a number to the second power is just multiplying the number by itself. Incidentally, if you start with the area of a square and want to find the length of its sides, the procedure is called "finding the square root."

Similarly, all the edges of a cube are the same, and consequently, you find the volume of the cube by "cubing," that is, raising the length of any edge to the third power.

It's also possible to raise expressions to "non-integer" powers, for example,

$$(j+7)^{2.5}.$$

This cannot be explained without the use of logarithms, which is a topic that is beyond what you need to understand this book. I will use this notation when I have to calculate, for example, 2.5 years' worth of interest on a loan. Your calculator or spreadsheet will handle this correctly; you don't need to worry about it.

1.6 SUMMATIONS

This topic is *not* necessary for understanding the rest of the book. It's an introduction, however, to a powerful notation that allows us to work with numbers from arbitrarily large lists in a very concise manner.

Looking at Table 1.3 again, suppose that I want to add up the values of e from e_5 to e_8. That is, add up $e_5 + e_6 + e_7 + e_8$. Just for the record, I don't know why I'd ever want to do this with a Life Table. There are cases, however, such as adding up the interest payments on a loan, where I often want to do this.

The shorthand notation involves the use of the uppercase Greek letter sigma:

$$\sum_{i=5}^{8} e_i.$$

How to read this: Beneath the sigma you see "$i = 5$." Above the sigma you see "8." This means that we want to add up all the terms to the right of the sigma for $i = 5, 6, 7,$ and 8:

$$\sum_{i=5}^{8} e_i = e_5 + e_6 + e_7 + e_8.$$

To finish the job, we have to look at the appropriate e values in the table:

$$\sum_{i=5}^{8} e_i = e_5 + e_6 + e_7 + e_8 = 70.8 + 69.8 + 68.9 + 67.9.$$

I won't bother actually adding these numbers up because I don't really care about the answer; I'm just showing how the notation works.

But why bother with such an esoteric notation just to show that I want to add up four numbers? The convenience is that I can represent huge sums of numbers concisely.

For example, Table 10.1 has 101 rows of numbers representing some information from age = 0 to age = 100. To show that I want to add up all the e values in this table, I write

$$\sum_{i=0}^{100} e_i.$$

This can be extended to more complicated expressions, such as

$$\sum_{i=5}^{8} (e_i + 7)^2 = (e_5 + 7)^2 + (e_6 + 7)^2 + \ldots$$

The variable i is sometimes called the "index." Note that it can also be used directly in the expression to be evaluated:

$$\sum_{i=5}^{8} i e_i = 5e_5 + 6e_6 + 7e_7 + 8e_8.$$

1.7 GRAPHS AND CHARTS

When looking at relationships between variables, the formula tells it all. Very often, however, a picture is indeed worth a thousand words in "giving us a feeling" for what the formula is telling us.

We will often be presented with a graph that we'll study to gain some insight into the information the graph is presenting. Conversely, we will often need to be able to create a graph to show a formula that we are interested in. I'll take this latter approach first.

Let's start with a simple formula:

$$y = 27,000 - 2,000x.$$

This formula gives us a value for the variable y when we give it a value for the variable x. These variables might stand for the depreciation of a car's value, the interest on a loan, the number of years that you will hold a loan, and so on.

Before I draw a graph, I have to know what values I will have to consider. Suppose that this formula tells us the value of a car over time: x will be the age of the car in years. The lowest value that x can possibly have is 0—that's the year when the car is brand new. There's no mathematical reason why x can't be less than 0 (negative numbers), but it doesn't make sense when I want x to stand for the age of the car. Let's say I only want to look at the first 10 years of the car's life. The largest value x can have is 10.

Next, I'll create a table showing y for various values of x. If possible, I recommend that you create this or a similar table on a computer spreadsheet. Since I don't know what spreadsheet program you're using, I can't give detailed directions for creating a graph from this table—but every spreadsheet program I've seen for the past 25 years has had graphing capability, so consult your manuals or do some searching online.

Table 1.4 Data for the First Graph Example

X	Y ($)
0	27,000
2	23,000
4	19,000
6	15,000
8	11,000
10	7,000

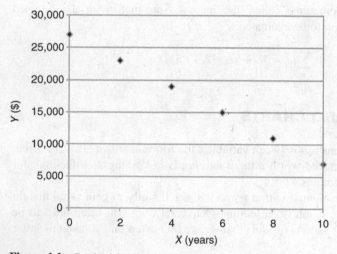

Figure 1.1 Graph of the data in Table 1.4.

As is seen in Table 1.4 and Figure 1.1, it's conventional to show the units for a variable in parenthesis after the variable, for example, X (years) in Figure 1.1.

Table 1.4 shows the data generated using the above formula, for even number values of x between 0 and 10. Why even number values? Here I have a choice. I can get very detailed ($x = 0, 0.2, 0.4, 0.6, \ldots, 9.8, 10$) or I can get very sparse ($x = 0, 5, 10$). There is no magic answer for what is the best thing to do. This is, unfortunately, as much an art as a science. You want enough detail so that the graph conveys all the details of the data, but you don't want to bury yourself (or your computer) in reams of numbers.

Figure 1.1 shows the graph for the data in Table 1.4. The horizontal line along the bottom (the horizontal axis) has the label "X" beneath it and shows numbered points between 0 and 10. The vertical line along the right (the vertical axis) has the label "Y" to the left of it and shows numbered points between 0 and 30,000.

Each diamond corresponds to a row in the table. If you draw a vertical line up from the number 4 on the horizontal axis and draw a horizontal line to the right from the number 18,000 on the vertical axis, there is a small diamond at the point where these lines cross.

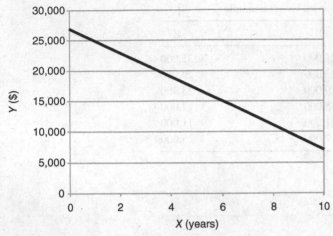

Figure 1.2 Continuous interpolation graph of data in Table 1.4.

Looking at the graph can quickly give you an idea of what's happening: the car value is dropping rapidly with passing years.

What about, say, $X = 5$? There is no diamond on the graph. I could have included $X = 5$ and the corresponding Y value in the table. Alternatively, once I'm sure that there are no "surprises" in the curve, I can draw a continuous curve rather than a discrete point, as is shown in Figure 1.2.

In Figure 1.2, I've taken the same data that I used for Figure 1.1 but I connected all the diamonds (and I'm not showing the diamonds). This is a very convenient way to show things—when you're very sure you know what's happening. Some formulas do funny things. Fortunately, we'll only be looking at fairly "well-behaved" formulas in the upcoming chapters, so there's no reason to dwell on mathematical curiosities. Connecting the dots between data points on a graph is called interpolation. Connecting these data points can sometimes lead to erroneous conclusions. Suppose the horizontal axis represents a number of apples and the vertical axis represents the cost of this number of apples in a store. The data might not be obvious because the store owner is giving discounts on large purchases. Connecting the points with a continuous line might give the impression that because you can estimate the cost of 1.5 apples, you can go into the store and buy 1.5 apples. In other words, connecting the points might give an incorrect impression that any value on the horizontal axis is possible.

When a graph describes how some variable such as the car value in the example above changes due to the change of another variable (the year in the example above), the graph is often entitled either "car values *as a function of* time," "car values versus time," or "car values with respect to time." This is just a shorthand jargon to tell you to expect a graph with car values on the vertical axis and time on the horizontal axis.

Graphs are especially useful for comparing two or more sets of data, that is, data that came from tables with three or more columns. The first column almost invariably becomes the horizontal axis on a graph.

Table 1.5   Data for the Second Graph Example.

X	Y	Z
0	27,000	35,000
2	23,000	29,000
4	19,000	23,000
6	15,000	17,000
8	11,000	11,000
10	7,000	5,000

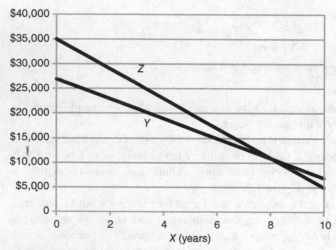

Figure 1.3   Graph of the data in Table 1.5.

Consider Table 1.5 and Figure 1.3. The table has three columns, labeled X, Y, and Z. Again considering car depreciation, Y and Z could represent the values of two different brands of cars, while X still represents the age of the car in years. In the figure, I've put the Y and Z labels inside the graph itself, each label nearest to the appropriate curve. The left (vertical) axis relates to both Y and Z.

Looking at Figure 1.3, it's pretty clear that car brand Z, while costing more when new, depreciates faster than car brand Y, and after about 8 years, brand Z is actually worth less than brand Y.

Another popular style of graph is the "histogram." The word histogram is derived from Greek roots and has to do with a drawing with information set upright. For practical purposes, it's another way to show some graphical information.

Figure 1.4 shows the same data as Figure 1.1—that is, the data of Table 1.4. As you can see, there is very little difference between these two figures. In the histogram version (Fig. 1.4), there is a solid bar reaching up from the horizontal axis to the place where Figure 1.4 had the diamond. Both graphs are read exactly the same way.

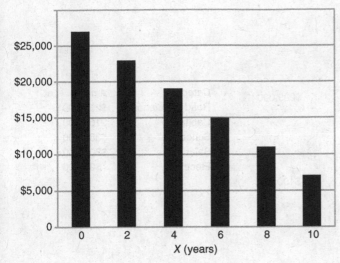

Figure 1.4   Histogram of the data in Table 1.1.

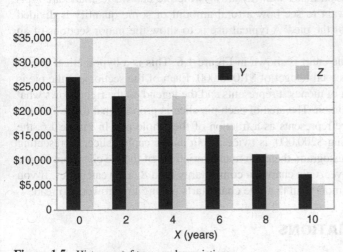

Figure 1.5   Histogram of two car depreciations.

Figure 1.5 shows the histogram version of Table 1.5, where I am comparing two sets of data. Note that in this case, I am explaining which set of bars belongs to which variable by means of little squares, appropriately labeled. This type of histogram can be extended to three or more variables, but things start getting very crowded and hard to grasp.

The histograms I've just shown aren't the only kinds of histograms; there are many variations on the theme. For the purposes of this book, however, these are enough.

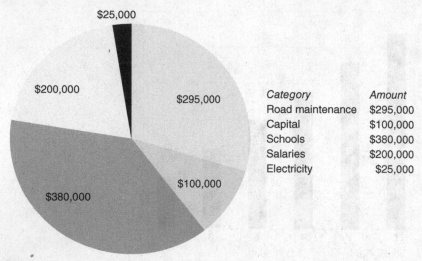

Category	Amount
Road maintenance	$295,000
Capital	$100,000
Schools	$380,000
Salaries	$200,000
Electricity	$25,000

Figure 1.6   Pie chart example.

The last type of graph that I want to show is the pie chart. Pie charts are typically used when you want to see how a total amount of some quantity is divided into pieces, or "slices of the pie." A typical use is to show the major sections of an organization's budget.

An example of a pie chart is shown in Figure 1.6. This is a hypothetical breakdown of a community's total budget of $1,000,000. Each of the sections of the pie is labeled with the amount of money it represents and the legend on the right shows what this money is being spent on. The area of each pie slice is proportional to the fraction of the total that the slice represents as a fraction of the whole pie. In Figure 1.6, the salaries slice, representing $200,000, is twice as big as the capital slice, representing $100,000. As with the histogram, there are many variations of the pie chart, but Figure 1.6 is fairly representative. A pie chart can comfortably hold 8 or 10 categories. If you try to squeeze in many more than that, the chart starts getting too crowded to read.

1.8   APPROXIMATIONS

We frequently don't need to know an answer to many decimal places. When we give someone directions to drive to our house, we usually say something like, "Get off the highway at exit 14, go right and follow the road for about 12 mi until you see an old church on the right."

We could have said "Follow the road for 11.87 mi" but "about 12" gives enough information to tell someone when to start looking for the church. I don't need to delve into the theory of approximations. Instead, I'll use some commonsense rules, such as "about 14 mi" means that the number is closer to 14 than it is to 13 or 15.

The mathematical expression

$$x \approx 14$$

means that "x is approximately equal to 14." Other ways of writing this are $x \sim 14$ and $x \cong 14$.

The number 2,123,774 has seven *significant figures*. It is approximately 2 million. I have to be careful, however, not to say that the number is approximately 2,000,000 because when I do this I am implying that I precisely know all the digits I put down: that is, that the last digit is indeed 0 and not 4.

All of the above issues are resolved by using a notation sometimes called "scientific" notation, which specifies exactly how many digits of a number are known and how many are just place keepers (zeros to the left of the decimal place). For the purposes of this book, I'll be careful just to use commonsense approximations and not worry about the mathematical implications.

Another name for approximating is *rounding*. When I know a currency amount to the nearest penny and I *round* it to the nearest dollar, I'm approximating the amount to two fewer significant figures than I started with. The rules are simple. If the amount after the decimal place is 50 or less, just drop this amount. If the amount after the decimal place is 51 or more, add 1 to the number of dollars and drop the amount after the decimal place. A few examples:

$151.23 rounded to the nearest dollar is $151.

$26.76 rounded to the nearest dollar is $27.

$315.00 rounded to the nearest dollar is $315.

You can also round to the nearest ten dollars, to the nearest million dollars, and so on. For example,

$514,676.26 rounded to the nearest hundred thousand dollars is $515 hundred thousand dollars—not $515,000.00 or $515,000.

Rounding to the nearest billion dollars is usually restricted to members of Congress.

1.9 RATES—AVERAGE AND INSTANTANEOUS

This section is useful for understanding the mathematics of the average and incremental IRS income tax rates discussed in Chapter 9. It is not necessary for understanding Chapter 9 or anything else in this book.

Suppose I were to take a walk for 25 minutes. I'm walking along a marked track, so I know exactly how far I've walked at all times. Every few seconds, I write down how long I've been walking and how far I've walked. After I've finished my walk, I produce the graph shown in Figure 1.7, interpolating my data as described above.

At the end of 25 minutes, I've walked about 2,200 feet. As the graph shows, I started off walking at a good pace and then I slowed down. From about 8 minutes to about 13 minutes, I hardly moved at all. Then I started walking faster and faster, until the end of my walk. The total distance traveled (2,200 feet) divided by the total time spent (25 minutes) is called the average rate of distance traveled. On this graph,

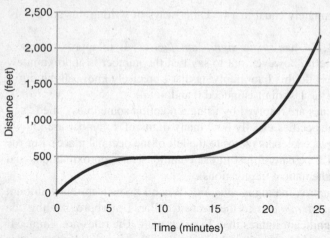

Figure 1.7 Graph of my walk—distance versus time.

it will be measured in feet per minute. "Rate" usually refers to something changing over time, so when I say "rate of distance traveled," the phrase "with respect to time" is implied. When the horizontal axis of a graph is something other than time, the definition of rate must be spelled out carefully.

The rate of distance with respect to time is such a common number that it has been given its own name, *speed*. If the distance is measured in feet and the time is measured in minutes (these are called the *units* of measurement), then the speed is measured in feet per minute. Speed can of course also be measured in miles per hour, inches per second, and so on. Since my speed was varying over the course of my walk, dividing the total distance traveled by the total time spent gives me the *average speed*:

$$\text{Average speed} = \frac{2{,}200 \text{ feet}}{25 \text{ minutes}} \approx 88 \frac{\text{feet}}{\text{minute}}.$$

If I draw a straight line connecting the start of my walk (time = 0, distance = 0) to the end of my walk (time = 25, distance ~ 2,200), this line represents how my walk would have been graphed had I walked at a constant speed, identically the average speed (Figure 1.8).

Mathematically, the property of a line describing the change in the vertical axis over the length of the line divided by the change in the horizontal axis over the length of the line is called the *slope* of the line. When a graph is showing distance on the vertical axis and time on the horizontal axis, the slope is the speed.

If I were to draw a line from my position at, say, time = 10 to time = 20, then the slope of this line would be my average walking speed between times 10 and 20.

Finally, I would like to be able to mathematically describe my speed at different times during the walk. Graphically, this means that I want to know the slopes of lines "tangent to" my graph at arbitrary points (just touching the graph at these

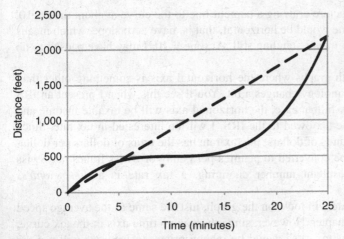

Figure 1.8 Graph of my walk showing average speed.

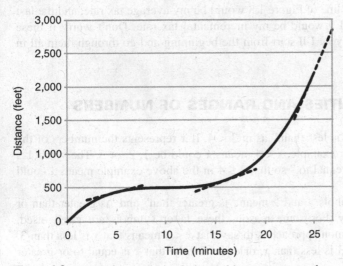

Figure 1.9 Graph of my walk showing several instantaneous speeds.

points). A few examples of this are shown in Figure 1.9. Mathematically, this is a topic in differential calculus, which we certainly aren't going to get into.

Fortunately, what I'm describing is very easy to picture intuitively. If I ask, for example, "How fast was I walking at time = 16?" then mathematically I am asking, "What is the slope of the line tangent to the curve at time = 16?"

In Figure 1.9, you see that I am asking about the line that's "bumping up" to and just touching the curve at time = 16. The answer is, at time = 16, I was walking 54 feet per minute. You can also see that I was walking a little slower than this at time = 5 and a good bit faster than this at time = 25.

Also, if you can picture drawing a tangent line to the curve at about time = 10, you will see that the line would be horizontal, that is, have zero slope, which means zero speed, which also means standing still. At time = 10, I may have paused to tie a shoelace.

When dealing with graphs where the horizontal axis is something other than time, some of the terminology changes a bit. You'll see this when I present an IRS tax curve in Chapter 9. In that case, the horizontal axis will be taxable income and the vertical axis will be tax owed to the IRS. I will be interested in tax rate. Since both of the axes have units of dollars, the tax rate has the units of dollars per dollar. Since these units can be converted to pennies per penny or Swiss francs per Swiss francs, without the resultant number changing, a tax rate is a *dimensionless* quantity.

The average tax rate will look, on the graph, just the same as the average speed on the graphs in this chapter. However, since there is no time axis in the tax curve, the term "instantaneous tax rate" would be inappropriate. Instead, we will use the term "incremental tax rate."

If Figures 1.7–1.9 represented an IRS tax curve instead of a description of my walk, then the straight line of Figure 1.8 would be my average tax rate, and the last dotted line of Figure 1.9 would be my incremental tax rate. Don't worry if these terms seem strange to you; I'll start from the beginning and go through them all in Chapter 9.

1.10   INEQUALITIES AND RANGES OF NUMBERS

The symbol < means "is less than," as in $3 < 4$. If x represents the numbers of the months of the year, for example, $x < 4$ means x could be 1, 2, or 3. The symbol $\leq$ means "is less than or equal to," so that $x \leq 4$ in the above example means x could be 1, 2, 3, or 4.

Similarly, the symbols > and $\geq$ mean "is greater than" and "is greater than or equal to," respectively. For some reason, these latter symbols are rarely used. Instead, the more common approach is to say that $x < 3$ means that x is less than 3, and $3 < x$ means that 3 is less than x, or equivalently, that x is equal to or greater than 3.

These symbols let us describe a range of numbers conveniently. For example,

$$3 < x < 7$$

means that x is somewhere between 3 and 7, but is not equal to either 3 or 7, while

$$3 < x \leq 7$$

means that x is somewhere between 3 and 7, possibly 7 itself, but not 3.

This notation will prove to be very useful in describing income tax brackets. For example, if y is your income, then the simple table

Income	Taxes
$y < \$10,000$	$25
$\$10,000 \leq y < 22,500$	$235
$\$22,500 \leq y$	$1,000

means that if your income is less than $10,000, your tax is $25; if your income is $10,000 or more but less than $22,500, your tax is $235; and if your income is $22,500 or more, your tax is $1,000. (This is not an actual tax table; I'm just interested in describing ranges of numbers using inequality signs here.)

PROBLEMS

1. Evaluate the following arithmetic expressions:
    (a) $7 - (12 - 5)$
    (b) $12(14 - 6)$
    (c) $\dfrac{16 - (3 + 7)}{3(7 - 5)}$
    (d) $(12 - 2)(7 + 3)$
    (e) $12 - 2(7 + 3)$
    (f) $(12 - 2)7 + 3$
    (g) $6.2 + 1/3$

2. Given that $x = 6$, $y = 2$, and $z = 3$, evaluate the following arithmetic expressions:
    (a) $x + y + z$
    (b) $z(x - 3)(y + 2)$
    (c) $\dfrac{x + 2}{y - 3} + 2.25$
        $\dfrac{}{z - 4}$
    (d) $x(x - 1)(x + 2)$

3. Refer to the table for the problem set 3:

T (hours)	P (\$)	N (number of wallets sold)
0	25.00	6
2	24.00	2
4	23.00	2
6	21.00	4
8	18.00	8
10	14.00	16

This table is for the price (P) of a wallet at a luggage counter over the course of a day set by a store owner who would be getting a shipment of new wallets the following day and wanted to make sure that he had cleared the old wallets out of his inventory before the new ones came in. T is the time in hours, starting the count from when the store opened. N is the number of wallets sold at the corresponding price (same row in the table).

(a) If we use subscripted notation to refer to the table entries for T and P, what is T_1, P_3, N_4, and T_5?

(b) How many wallets were sold at more than $20 per wallet? How many wallets were sold all together?

(c) If the storekeeper paid $12 each for these wallets, not counting overhead, did he make or lose money that day? If overhead (rent, electricity, etc.) costs $100 a day, did the storekeeper make or lose money that day?

4. Let's extend problem 3 to a 2-day rather than a 1-day wallet sale. On the second day, the storekeeper decides to repeat his change-the-price-every-two-hours strategy, but he drops each price by 10% from its amount on the first day. His sales in each 2-hour period on the second day are 50% higher than they were on the first day.

(a) Create a table just like the table in problem 3, except use numbers for the second day.

(b) Using the same assumption about overhead as in problem 3, how much money did the storekeeper make or lose on the second day?

5. Sketch a histogram based on the table of problem 3 using T as the horizontal axis and N as the vertical axis.

6. The table below shows a business plan to put items on sale. Shown are the original price and the sale price of a list of items. Calculate the percent change of all the items to the nearest percent and put these numbers in the empty percent change locations in the table. Calculate the percent sale numbers (% Sale = $100 - $ % Change). Then plot a graph of these data, placing the original price on the horizontal axis and the percent sale on the vertical axis.

Original price ($)	Sale price ($)	% Change	% Sale
289.99	217.49		
249.99	199.99		
127.50	102.00		
99.99	84.99		
59.79	53.99		
37.50	33.75		

7. For $x = 0.5$ and then again for $x = 1.2$, fill out the table below:

n	$(0.25 + x)^n$ $x = 0.5$	$(0.25 + x)^n$ $x = 1.2$
0		
1		
2		
3		
4		

(Use a calculator or a spreadsheet for this problem; doing it by hand is a very tedious job.) Can you draw any conclusions about numbers raised to the 0th power and numbers raised to the 1st power?

8. Approximate (round) the following numbers to the nearest dollar: $12.87, $22.22, $53.50, and $1,719.88.

9. The following table describes a walk I recently took. I started at 1:30 and walked until 3:30:

Time walked (miles)	Distance
1:30	0
1:45	0.25
2:00	0.5
2:15	0.75
2:30	1.00
2:45	1.50
3:00	2.00
3:15	2.50
3:30	3.00

Draw a graph using time walking (in hours) as the horizontal axis and distance walked (in miles) as the vertical axis. What was my average speed for the entire walk? (Don't forget to specify your units.) Can you estimate the instantaneous speeds near the beginning of the walk and near the end of the walk?

Chapter 2

Compound Interest

The most common, if not universal, way to express the amount of interest to be paid on a loan is the *annual percentage rate* (APR). The interest is expressed as a percentage or a fraction of the amount of money loaned if the money were to be loaned, with no intermediate payments or corrections, for a year.

Calculating interest is very simple. An important point to remember is that while interest is usually expressed as a percentage, for example, "6% per year," calculations must always use the decimal or fractional equivalent of this percentage:

$$6\% = \frac{6}{100} = 0.06.$$

The interest due after a year on a $1,200 loan, for example, is then

$$\text{Interest} = (\$1,200.00)(0.06) = \$72.00.$$

A type of interest calculation that is rarely used is called *simple interest*. In a simple interest calculation, the interest is equal to the amount borrowed, or the *principal*, times the rate of interest, times the amount of time the money is borrowed. The formula is

$$\text{Interest} = (\text{Principal})(\text{Rate})(\text{Time}).$$

Time must be expressed in the same units as rate. In other words, if the rate is expressed in percentage per year, then the time must also be expressed in years. If the rate is in percentage per month, then the time must be in months. The interest calculated, of course, is for the time period that the rate is expressed in.

The same example as above would therefore be written as

$$\text{Interest} = (\$1,200.00)(0.06)(1.0) = \$72.00.$$

The following are a few more examples:

The simple interest on an $800 loan with an interest rate of 8% per year after 2 years is

$$\text{Interest} = (\$800.00)(0.08)(2) = \$128.00.$$

Understanding the Mathematics of Personal Finance: An Introduction to Financial Literacy, by Lawrence N. Dworsky
Copyright © 2009 John Wiley & Sons, Inc.

The simple interest on a $2,500 loan with an interest rate of 5% per year after 18 months is (remember that 18 months is 1.5 years)

$$\text{Interest} = (\$2,500.00)(0.05)(1.5) = \$187.50.$$

As I have stated above, simple interest is rarely used in real transactions. For that matter, the term *interest* when used alone does not mean *simple interest*. It means *compound interest*.

In a compound interest loan, there is a period of time called the *compounding interval* or the *compounding period*. Suppose you are told that your loan will be compounded monthly, starting 1 month after you originate the loan. Starting 1 month after you originate the loan, and monthly thereafter, interest is calculated on the total amount of money that you owe. This is very different from simple interest in that the interest from each compounding calculation is added to the amount of money you owe, now called your *balance*, and subsequent interest calculations are calculated based on this balance.

I need to introduce one other convention before presenting some interest calculation examples. It is conventional to present the terms of a loan as the annual interest rate and the compounding intervals. The actual interest rate used to calculate the interest is just the APR divided by the compounding interval (remember to put both of these numbers into the same units).

Examples:

1. If interest is compounded monthly, then it is compounded 12 times a year. If a 6% APR loan is compounded monthly, then each month the interest is 6%/12 = 0.5% of the balance of the loan.

2. Interest compounded quarterly is compounded 4 times a year. A 10% loan compounded quarterly earns 10%/4 = 2.5% interest every quarter (every 3 months).[1]

3. Interest compounded annually is compounded once a year. In this case, the stated APR is identically the interest rate used for calculation.

Now let's look at some real compound interest loan calculations. I'll start by explaining the different columns in the tables that detail these examples.

Consider a $5,000.00 loan is taken at 5% interest, compounded annually.

In Table 2.1, the first column is a list of compounding periods. The loan is taken at a starting compounding period of 0, with compounding periods occurring annually afterwards. The second column in Table 2.1 shows actual dates. In this example, I assumed that the loan was taken on July 14, 2007, and that it will be paid back after 15 years, on July 14, 2022.

Also shown in Table 2.1 is a third column labeled *interest* and a fourth column labeled *balance*. Since the loan was originated on July 14, 2007, at that date there have been no interest compoundings yet. The interest column therefore shows $0, and the balance (how much you owe) shows the principal, $5,000.00.

---

[1] The term "10% loan," unless expressly explained otherwise, is synonymous with a "10% APR loan."

Table 2.1   A Compound Interest Balance Sheet

Compounding interval	Date	Interest ($)	Balance ($)
0	July 14, 2007	0.00	5,000.00
1	July 14, 2008	250.00	5,250.00
2	July 14, 2009	262.50	5,512.50
3	July 14, 2010	275.63	5,788.13
4	July 14, 2011	289.41	6,077.53
5	July 14, 2012	303.88	6,381.41
6	July 14, 2013	319.07	6,700.48
7	July 14, 2014	335.02	7,035.50
8	July 14, 2015	351.78	7,387.28
9	July 14, 2016	369.36	7,756.64
10	July 14, 2017	387.83	8,144.47
11	July 14, 2018	407.22	8,551.70
12	July 14, 2019	427.58	8,979.28
13	July 14, 2020	448.96	9,428.25
14	July 14, 2021	471.41	9,899.66
15	July 14, 2022	494.98	10,394.64

The first interest calculation is performed on July 14, 2008. Since this is the first such calculation, column 1 shows this as compounding interval #1. Five percent APR compounded annually is just 5% interest per calculation. This interest is therefore

$$\text{Interest} = (\$5,000.00)(0.05) = \$250.00.$$

The balance, that is, how much you owe, is now the previous balance ($5,000.00) plus this interest ($250.00), which is $5,250.00.

July 14, 2009, is the date of the second compound interest calculation. Here's where compound interest is very different from simple interest. The interest calculation is based not on the principal ($5,000.00) but on the balance at the time of the calculation, in this case $5,250.00:

$$\text{Interest} = (\$5,250.00)(0.05) = \$262.50.$$

Notice that this interest value is higher than the first interest value ($262.50 vs. $250.00), and that the balance grows to $5,612.50.

I'll repeat this calculation in detail one more time. On July 14, 2010, the third compound interest calculation is

$$\text{Interest} = (\$5,512.50)(0.05) = \$275.63,$$

and the new balance is $5,788.13.

Each time the calculation is performed, the interest is larger. This is sometimes called the "miracle of compound interest." If these calculations are for the $5,000

Table 2.2   A Compound Interest Balance Sheet for 2 Years of a $10,000 Loan with a 10% Interest Rate, Compounded Monthly

Compounding interval #	Interest ($)	Balance ($)
0	0.00	10,000.00
1	83.33	10,083.33
2	84.03	10,167.36
3	84.73	10,252.09
4	85.43	10,337.52
5	86.15	10,423.67
6	86.86	10,510.53
7	87.59	10,598.12
8	88.32	10,686.44
9	89.05	10,775.49
10	89.80	10,865.29
11	90.54	10,955.83
12	91.30	11,047.13
13	92.06	11,139.19
14	92.83	11,232.02
15	93.60	11,325.62
16	94.38	11,420.00
17	95.17	11,515.16
18	95.96	11,611.12
19	96.76	11,707.88
20	97.57	11,805.45
21	98.38	11,903.83
22	99.20	12,003.03
23	100.03	12,103.05
24	100.86	12,203.91

deposit you put into a savings bank, then you can see how the interest takes on a life of its own and, over the years, earns you more money than the principal you started with. After 15 years, the balance is slightly more than twice the original deposit (principal).

What has been presented so far is really the total subject of compound interest, and the basis for almost everything else in this book. What has to be added to this chapter are several practical cases that come up frequently, some formulas for working with these cases efficiently, and techniques for "running the numbers" yourself.

Table 2.2 shows the calculated interest and balance over time for a $1,000.00 loan with a 10% annual interest rate, compounded monthly for 2 years (24 months). Since there are 12 months in a year, each month's interest is calculated based on a monthly rate of

$$i = \frac{10\%}{12} = 8.333\%.^2$$

Since only the elapsed time since the origination of the loan matters for interest calculations, not the actual date of the loan, I omitted the actual dates in this table.

One last reminder: When performing the division shown above, make sure that units agree. If the interest rate is percentage per year, then the compounding interval must be the number of compounding intervals per year, and so on.

The calculations required to produce Table 2.2 are the same as those used to produce Table 2.1, so I won't repeat them here.

Table 2.2 shows that after 1 year (12 months), the balance is $11,047.13. Isn't this a bit odd? If I borrow $10,000 at an APR of 10%, shouldn't I just owe $10,000 + (0.1)($10,000) = $11,000 at the end of a year? This discrepancy is a result of the way things are quoted. Because the loan is compounded many times over the course of the year, the "miracle of compound interest" causes the balance to grow to a larger number than we'd see if we just compounded once at the end of the year. In order to owe $11,047.13 if the balance were compounded only once a year, the interest rate would have had to be 10.4713% (work it out; it's pretty straightforward). This latter number is referred to as the EAPR, or effective annual percentage rate.[3]

2.1 SOME MATHEMATICS

You can skip this section if you aren't interested in all the details of the calculations. I recommend that you at least glance through this section to try and get an idea of what's going on in various calculations.

At the outset of the loan, the interest is just the principal times the interest rate per compounding interval, therefore we may write

$$\text{Interest} = P\frac{R}{n},$$

where P is the principal, R is the interest rate per year, and n is the number of compounding intervals per year.

To get the new balance, you add this interest to the principal:

$$\text{Balance} = P + P\frac{R}{y}.$$

You can see that the principal P appears twice in this equation. The rules of algebra let us write this same formula as

$$\text{Balance} = P\left(1 + \frac{R}{y}\right).$$

---

[2] In this example, 8.333% is an approximation, because of a "repeating decimal."

[3] When there is a possibility for confusion, the APR is sometimes called the NAPR, or nominal annual percentage rate, to emphasize the distinction between it and the EAPR.

Now, suppose you want to get the balance after the second compounding period. You just repeat what you've done, but instead of using the principal, you use the last balance. The new balance is then

$$\text{Balance} = P\left(1+\frac{R}{y}\right)\left(1+\frac{R}{y}\right).$$

After three compounding periods, the balance would be

$$\text{Balance} = P\left(1+\frac{R}{y}\right)\left(1+\frac{R}{y}\right)\left(1+\frac{R}{y}\right)$$

and so on.

Using exponential notation,

$$\text{Balance} = P\left(1+\frac{R}{y}\right)^{n},$$

where n is the number of compounding intervals.

This formula is useful for calculating the balance after any number of compounding intervals on a pocket calculator because all but the simplest of pocket calculators will have the ability to raise an expression to the power n, as shown.

2.2 MY WEBSITE SPREADSHEET

If you'd like to try my website spreadsheet calculators, please take a moment and glance through Chapter 15. You'll find instructions for obtaining a free spreadsheet program if you need one, as well as instructions for downloading and using my spreadsheets.

Tables 2.1 and 2.2 were generated with my spreadsheet Ch2CompoundInterest. xls. After opening the spreadsheet, click on the Basic tab. Table 2.3 shows part of Table 2.2 with the spreadsheet's row and column designations added.

Table 2.3   Table 2.2 (First Few Entries Only Are Shown) in Spreadsheet Format

	D	F	G
	Compounding interval	Interest ($)	Balance ($)
1	0	0.00	10,000.00
2	1	83.33	10,083.33
3	2	84.03	10,167.36
4	3	84.73	10,252.09
5	4	85.43	10,337.52
6	5	86.15	10,423.67

The first example above (Table 2.1) is generated using my spreadsheet with the following entries (to the left of the green line):

Nr Years = 20	The number of years that interest is accruing
Cmpds per Year = 1	The number of compounding intervals per year
Principal = $5,000.00	
Rate = 5.00%	This is the APR
Up-front = $0.00	Up-front fee. This hasn't been discussed yet. Set it = 0 for now.

For the second example above (Table 2.2):

Nr Years = 2	The number of years that interest is accruing
Cmpds per Year = 12	Monthly compounding
Principal = $10,000.00	
Rate = 10.00%	APR
Up-front = $0.00	

Your results should agree with Table 2.2. The spreadsheet (to the right of the green line) should look exactly like Table 2.3 (the first few lines of the spreadsheet, that is).

2.3   ONLINE CALCULATORS

I found so many excellent compound interest calculators that I'm only listing a representative sample. All of the calculators I found have a good graphical interface. Most of them want the interest rate stated as a percentage, that is, 10%, not 0.1. Some of them let you invert the problem arbitrarily. That is, you can pick any three of the four variables' future value, principal, years, and rate, and the calculator will solve for the fourth variable. Future value in the examples above is the balance at the end of the loan:

1. www.moneychimp.com/calculator/compound_interest_calculator.htm;
2. www.1728.com/compint.htm;
3. www.webmath.com/compinterest.html.

2.4   SCALING

This section is not necessary for understanding and working with the rest of the book. Its conclusion, however, is very easy to understand, so glance at it even if you're uncomfortable with the math.

Looking at the formula for loan balance above, we see that there are four numbers necessary to calculate a balance: P, R, y, and n. Suppose we have done all our calculations and we want to know what happens if we were to borrow $10,000 instead of $5,000. In other words, we're doubling P.

For those of you comfortable with the algebra,

$$(2P)\left(1+\frac{R}{y}\right)^n = 2\left[P\left(1+\frac{R}{y}\right)^n\right] = 2b,$$

where b was the original balance.

What this is saying is that if I've calculated a balance for some numbers P, R, y, and n and then I double P (the principal), the balance doubles. If I were to divide the principal by a factor of 3, the balance would divide by a factor of 3. This property is called "linear scaling of the balance with respect to the principal."

Let's look at this situation without the algebra. If my friend and I open identical savings accounts and put in the same amount of money, at any time in the future we would expect our balances to also be identical. This means that starting with twice the principal must lead to twice the balance at any time.

This property of linear scaling *only* holds for the principal; if I change R, y, and/or n, the balance certainly will change, but there's no way to predict the change other than by actually doing the calculation.

2.5 PRORATION—WORKING INSIDE A COMPOUNDING INTERVAL

Suppose I want to pay off a loan 4 months and 20 days after taking the loan.[4] The bank isn't going to just use the results of the fourth compounding interval calculation and give me the interest they earned in the last 20 days. On the other hand, I don't think it's fair for the bank to use the fifth compounding interval calculation and charge me 10 days' worth of interest that the bank hasn't earned.

Let's say that there are 30 days in the current month. Take the interest rate per month $\left(\frac{R}{y}\right)$ and divide it by 30, giving us an interest rate per day, then we multiply this by the number of days, giving us an effective interest rate for the 20 days of the month:

$$eff = \frac{20}{30}\left(\frac{R}{y}\right) = \frac{20}{30}\left(\frac{0.1}{12}\right) = 0.00556.$$

The interest is just the above number times the balance after 4 months. The balance as of 20 days after the fourth payment is therefore $10,337.52 + 20(0.000278) ($10,337.52) = $10,394.95.

---

[4] I'm assuming that there are no prepayment penalties associated with this loan. This isn't always the case. How these penalties are often calculated will be shown in a later chapter.

The procedure is referred to as *prorating*; you convert the monthly interest rate to an effective daily rate and then calculate the interest earned in the appropriate number of days.[5]

On my spreadsheet Ch2CompountInterest.xls, click on the Prorate tab. To the left of the green line you'll see three new entries: number of days in month, day of proration, and payoff month number. For the example above, using the same $10,000 loan as in previous examples, enter 30, 20, and 4 for these variables, respectively.

To the right of the green line you'll see the same columns as in the Basic tab. On the top row, you'll see the daily interest rate calculated. Then, look at the new column, payoff. This column shows the payoff number for the twentieth day of a 30-day month after the fourth payment.

2.6 INITIAL CHARGES AND EFFECTIVE INTEREST RATE

Very often, a bank or loan company will charge some sort of loan initiation fee at the outset of the loan (I call these up-front costs). Suppose, still using Table 2.2 as our example, that the bank wants $100 up front for setting up the loan. One way to handle this would be for the bank to just give you $9,900 while recording the loan as $10,000; then Table 2.2 would still be correct.

Usually, however, this is not what happens. When you take a $10,000 loan, you probably want to walk out of the bank with $10,000 for whatever your purpose is. In this case, the bank adds the extra $100 into the loan. That is, it pretends that it really loaned you $10,100. Columns 2 and 3 of Table 2.4 shows the new balance worksheet for this loan. Naturally, at the end of 2 years when you go to repay the loan, you owe more than Table 2.2 predicts.

In the last two columns, I reset the principal to $10,000 and then *adjusted* the interest rate until the balance after 2 years was approximately the same as the balance at the same time in columns 2 and 3. I had to set the interest rate to 10.50% to get this result. As you can see, the loan with the $100 initial charge may be thought of as the same loan without this charge, but with an *effective* interest rate that's a bit higher than the originally stated interest rate.

Two points here: First, you can see the value of using a spreadsheet for these calculations. I was able to adjust the input parameters quite easily until I got the results I was looking for, using the Basic tab of my spreadsheet. If I wanted to be a little fancier, I could have set up both examples side by side—that's actually what I did to create Table 2.4.

Second, this *effective* interest rate calculation is a very good way to compare loans from two different lenders. Suppose you are offered two loans with the same number of payments but at different interest rates and with different initial charges. Which one is the better loan? By breaking the two loans down to the money you're actually borrowing and an effective interest rate, you can see which deal is better.

[5] This is actually charging simple interest for prorations inside one compounding interval.

Table 2.4   The Same Loan as Shown in Table 2.2 but with a $100 Initial Charge

Compounding interval #	Initial charge as charge		Initial charge as effective interest	
	Interest ($)	Balance ($)	Interest ($)	Balance ($)
0	0.00	10,100.00	0.00	10,000.00
1	84.17	10,184.17	87.50	10,087.50
2	84.87	10,269.03	88.27	10,175.77
3	85.58	10,354.61	89.04	10,264.80
4	86.29	10,440.90	89.82	10,354.62
5	87.01	10,527.91	90.60	10,445.22
6	87.73	10,615.64	91.40	10,536.62
7	88.46	10,704.10	92.20	10,628.81
8	89.20	10,793.30	93.00	10,721.82
9	89.94	10,883.25	93.82	10,815.63
10	90.69	10,973.94	94.64	10,910.27
11	91.45	11,065.39	95.46	11,005.73
12	92.21	11,157.60	96.30	11,102.03
13	92.98	11,250.58	97.14	11,199.18
14	93.75	11,344.34	97.99	11,297.17
15	94.54	11,438.87	98.85	11,396.02
16	95.32	11,534.20	99.72	11,495.74
17	96.12	11,630.32	100.59	11,596.32
18	96.92	11,727.23	101.47	11,697.79
19	97.73	11,824.96	102.36	11,800.15
20	98.54	11,923.50	103.25	11,903.40
21	99.36	12,022.87	104.15	12,007.55
22	100.19	12,123.06	105.07	12,112.62
23	101.03	12,224.08	105.99	12,218.60
24	101.87	12,325.95	106.91	12,325.52

Since both loans have the same number of payments, it is the loan with the lower effective interest rate.

2.7   IN THE LIMIT—CONTINUOUS COMPOUNDING

This section just shows a point that is probably interesting for those comfortable with the math. It's not a necessary section. However, I recommend looking at the graph and reading through the description of the axes.

The balance of a loan that's compounded once a year at, say, 10% annual interest grows by 10% a year. As I have shown above, if the same loan is compounded monthly (12 times a year), the balance grows by 10.47% a year. What if the loan is compounded more often—weekly, daily, or even hourly? Does the effective interest rate just keep growing?

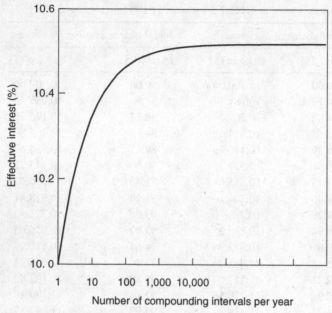

Figure 2.1   Effective interest rate versus the number of compounding intervals per year.

The answer to this question is shown in Figure 2.1. When 10% annual interest is compounded just once a year, the balance grows by exactly 10%. When it's compounded 12 times a year (monthly), the balance grows by about 10.45%. When it's compounded 100 or 1,000 or 10,000 or more times a year, the balance grows by about 10.52%. Once you're compounding more often than weekly, the EAPR stops growing by any meaningful amount.

The format of Figure 2.1 is a bit different from anything I've shown before. The horizontal axis doesn't start at 0 and the tic marks don't represent equal size steps (a "linear" axis). This graph is known as a "semilogarithmic" graph. (Don't let the name scare you.) In this graph, every tic mark on the horizontal axis represents a factor of 10 larger than the previous tic mark. If I didn't do this, I would have to choose between just showing what's happening for the Nr of Compounding Intervals Per Year varying between 1 and about 20—and therefore not showing how the curve "flattens out," or showing it between, say, 1 and 1,000 so I could show the flattening, and thereby squeezing the early growth details so tightly to the vertical axis that they would be unintelligible.

As to where the data for this graph came from, if you have a bit of a calculus background, you'll recognize that what I'm doing is just letting y approach infinity in the expression

$$EAPR = \lim_{y \to \infty} \left(1 + \frac{R}{y} \right)^y - 1 = e^R - 1,$$

where e is the base of natural logarithms, 2.718....

PROBLEMS

1. For a simple interest loan, calculate the following:
   - (a) The interest on a 1-year loan of $6,700 at 10% interest per year.
   - (b) The interest on a 3-year loan of $500 at 6% interest per year.
   - (c) The interest on a 7-month loan of $1,000 at 8% interest per year.
   - (d) The annual interest rate on a loan of $12,000.00 that paid back $15,600.00 after 3 years.

2. The APR of a compound interest loan = 18.0%. What is the compounding rate if the interest is compounded once, twice, three times, or four times a year?

3. A loan with an APR of 7.5% is compounded annually. If $1,250.00 is borrowed, find the balance each year for 3 years.

4. I took a loan for $10,000 that is accruing interest at an APR of 7.5%, compounded monthly. What is the balance every month for 18 months?

5. For the same loan as above, I decide to pay the loan back 13 days after my eleventh payment, in a 28-day month. What is my balance on the payoff day?

6. For the loan of problem 4, assume that my lender wanted an up-front fee of $250 to set up the loan. I want the lender to fold this fee into the loan. First, repeat problem 4 showing this up-front fee as additional principal. Second, return the loan to the original principal but show the effective interest rate due to the higher payments.

7. For the calculations in this problem, assume that the interest is compounded annually. Suppose you are able to take a 10-year loan at the loan rates shown in the table, and you are able to save the money for 10 years at the savings rates shown in the table. You borrow $10,000 and put it all into the savings account. Fill out the table. (This is known as working with "other people's money.")

Loan rate (%)	10-year balance ($)	Savings rate (%)	10-year balance	Profit
5	16,289	6		
5	16,289	7		
5	16,289	8		
6	17,908	7		
6	17,908	8		
7	19,672	8		

Chapter 3

Loan Amortization and Savings

3.1 LOANS

Many, if not most, loans are repaid with periodic payments. You make monthly mortgage payments, monthly car loan payments, and so on. Although it's not mathematically necessary, it's typical for the payment intervals and dates (how often and when you make payments) to coincide with the interest compounding intervals and dates. Keep in mind that the interest is calculated on the money you have owed for the past month (assuming monthly compounding) and does not reflect the payment you're about to make.

In some cases, your payments are credited on the day that they are received, and accrued interest is calculated by the day. Credit card companies (as discussed in Chapter 6) do this. I'm not going to explore this situation here because it doesn't really change the numbers very much while, at the same time, it does complicate things enough to make the understanding of the basic factors involved annoyingly more difficult. If you're having trouble getting statements from a lender to exactly match your own calculations based on this chapter, take a close look at your statements and see whether this situation is causing the problem.

Table 3.1 shows a $30,000 loan taken at an annual rate of 8%, compounded monthly for 4 years. The interest earned after the first month of this loan is $200.00, as shown. I didn't show all of the rows of the table, but I think the message is clear.

Table 3.2 is a repeat of the first few rows of Table 3.1 but with a payment column added. Assume that each month, on the same day that the interest is calculated, you make a payment of $200.00. Since this is exactly the amount of the first month's interest, the balance never changes, as shown in the table.

Paying exactly the month's accrued interest and thereby freezing the balance is sometimes called *making interest-only payments* on a loan. It's useful for loans that you intend to pay off in full soon. For example, if you buy a new house before you

Understanding the Mathematics of Personal Finance: An Introduction to Financial Literacy, by Lawrence N. Dworsky
Copyright © 2009 John Wiley & Sons, Inc.

Table 3.1 Amortization Table for a $30,000 Loan with 8% Annual Interest, Compounded Monthly for 4 Years

Compounding interval	Interest ($)	Balance ($)
0	0.00	30,000.00
1	200.00	30,200.00
2	201.33	30,401.33
3	202.68	30,604.01
4	204.03	30,808.04
5	205.39	31,013.42
6	206.76	31,220.18
7	208.13	31,428.31
42	262.63	39,657.03
43	264.38	39,921.41
44	266.14	40,187.55
45	267.92	40,455.47
46	269.70	40,725.17
47	271.50	40,996.67
48	273.31	41,269.98

Table 3.2 Amortization Table for a $30,000 Loan with 8% Annual Interest, Compounded Monthly for 4 Years and $200 a Month Payments

Compounding interval	Interest ($)	Payment ($)	Balance ($)
0	0.00	0.00	30,000.00
1	200.00	200.00	30,000.00
2	200.00	200.00	30,000.00
3	200.00	200.00	30,000.00
4	200.00	200.00	30,000.00
5	200.00	200.00	30,000.00

sell your old house, you might just pay the interest on the mortgage taken on the new house to keep your monthly expenses as low as possible while not letting your balance grow. As soon as your old house sells, you use the proceeds of the sale to pay off the loan.

If you make a monthly payment of more than the (first month's) interest, the balance will decrease. Table 3.3 shows this happening when a $300 payment is made each month. As you can see, both the monthly interest and the balance are decreasing with time. You are now *paying off the loan*. It is not necessary for all the payments to be equal to pay off a loan. As long as each payment is larger than the interest accrued since the last payment, the balance will decrease.

Table 3.3   Amortization Table for a $30,000 Loan with 8% Annual Interest, Compounded Monthly for 4 Years and $300 a Month Payments

Compounding interval	Interest ($)	Payment ($)	Balance ($)
0	0.00	0.00	30,000.00
1	200.00	300.00	29,900.00
2	199.33	300.00	29,799.33
3	198.66	300.00	29,698.00
4	197.99	300.00	29,595.98
5	197.31	300.00	29,493.29

Table 3.4   Amortization Table for a $30,000 Loan with 8% Annual Interest, Compounded Monthly for 4 Years with $732.09 a Month Payments

Compounding interval	Interest ($)	Payment ($)	Balance ($)
0	0.00	0.00	30,000.00
1	200.00	732.39	29,467.61
2	196.45	732.39	28,931.67
3	192.88	732.39	28,392.16
4	189.28	732.39	27,849.05
5	185.66	732.39	27,302.32
42	33.28	732.39	4,293.48
43	28.62	732.39	3,589.71
44	23.93	732.39	2,881.25
45	19.21	732.39	2,168.07
46	14.45	732.39	1,450.13
47	9.67	732.39	727.41
48	4.85	732.26	0.00

Table 3.4 shows what happens when the regular monthly payment is $732.39. I'm just showing the first few and the last few rows of the table. The last row of the table shows the point of this example. The regular monthly payment of $732.39 brings the balance to zero at the end of 48 months. The loan has been paid off in 48 months by a regular monthly payment of $732.39. This procedure is called *amortizing the loan*, and Table 3.4 is called an *amortization table*.

Before discussing about how I calculated the $732.39 payment, let me show some other results that occur when you pay off a loan with regular payments. First, look at the interest amounts corresponding to the first and the last payments. The first interest amount is $200.00; the last interest amount is only $4.85. This means that at the first payment, the balance is reduced by $732.39 − $200.00 = $532.39, while at the last payment, the balance is reduced by $732.39 − $4.85 = $727.54.

The calculated interest starts high and decreases with each payment, while the amount that the balance is reduced starts low and increases with each payment. This is a basic characteristic of how compound interest is calculated and the loan amortization process. It has nothing to do with a bank "front-loading the interest to get its profit out quickly" or anything of the sort. Early in the amortization of a loan, you still owe most of the principal, so the interest for each payment period is high. As you approach paying off the loan, you don't owe very much money, so the interest for each payment period is low.

Let's look at another example where these factors are much more apparent. Consider a $300,000 loan, at 8% annual interest compounded monthly, paid back with regular monthly payments over a 20-year period. This could be a mortgage on a home. I still haven't shown you how I came up with the payment amount but just by looking at the amortization table, you can see that it's correct. Table 3.5 shows this amortization table. I've added a few columns to the table that I'll discuss soon. Note also that I've replaced the title *compounding interval* with the title *payment #*. In this example, they're interchangeable.

Figure 3.1 shows the interest and the amount of each payment going to reduce the balance over the 20-year life of the loan. As you can see, not until a little past halfway through the 20-year life of the loan does the monthly balance reduction get greater than the monthly interest payment.

The second and third columns of Table 3.5 show some dates, starting with July 2008 (7/08). I assumed that the loan was taken on July 1, and the first payment was due on August 1, 2008. In the last column on the right, I add up all the interest payments

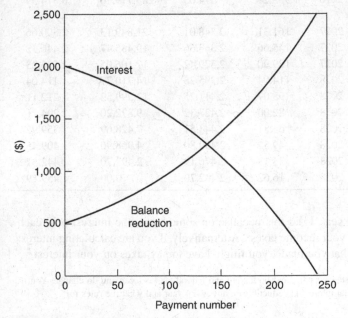

Figure 3.1 Example of payment allocation to interest and balance reduction in a regular payment loan.

Table 3.5 Amortization Table for a $300,000 Loan with 8% Annual Interest, Compounded Monthly for 20 Years with $2,509.32 a Month Payments

Payment #	Month	Year	Interest ($)	Balance reduction ($)	Balance ($)	Total interest per year ($)
0	7	2008	0.00	0.00	300,000.00	0.00
1	8	2008	2,000.00	509.32	299,490.68	2,000.00
2	9	2008	1,996.60	512.72	298,977.96	3,996.60
3	10	2008	1,993.19	516.13	298,461.83	5,989.79
4	11	2008	1,989.75	519.57	297,942.26	7,979.54
5	12	2008	1,986.28	523.04	297,419.22	9,965.82
6	1	2009	1,982.79	526.53	296,892.69	1,982.79
7	2	2009	1,979.28	530.04	296,362.66	3,962.08
8	3	2009	1,975.75	533.57	295,829.09	5,937.83
9	4	2009	1,972.19	537.13	295,291.96	7,910.02
10	5	2009	1,968.61	540.71	294,751.25	9,878.64
11	6	2009	1,965.01	544.31	294,206.94	11,843.65
12	7	2009	1,961.38	547.94	293,659.00	13,805.03
13	8	2009	1,957.73	551.59	293,107.41	15,762.75
14	9	2009	1,954.05	555.27	292,552.14	17,716.80
15	10	2009	1,950.35	558.97	291,993.16	19,667.15
16	11	2009	1,946.62	562.70	291,430.47	21,613.77
17	12	2009	1,942.87	566.45	290,864.01	23,556.64
18	1	2010	1,939.09	570.23	290,293.79	1,939.09
19	2	2010	1,935.29	574.03	289,719.76	3,874.39
231	10	2027	161.31	2,348.01	21,849.13	2,300.66
232	11	2027	145.66	2,363.66	19,485.47	2,446.32
233	12	2027	129.90	2,379.42	17,106.05	2,576.23
234	1	2028	114.04	2,395.28	14,710.77	114.04
235	2	2028	98.07	2,411.25	12,299.52	212.11
236	3	2028	82.00	2,427.32	9,872.20	294.11
237	4	2028	65.81	2,443.51	7,428.69	359.92
238	5	2028	49.52	2,459.80	4,968.90	409.45
239	6	2028	33.13	2,476.19	2,492.70	442.57
240	7	2028	16.62	2,492.70	0.00	459.19

made in each calendar year. I did this because on some loans, the interest paid each year is deductible from your income taxes.[1] Alternatively, if you are calculating interest paid to you for a loan that you made, you might have to pay taxes on your interest.[2]

[1] State tax laws vary from state to state, and both state and federal tax laws can and do change. You'll have to find out yourself what interest is deductible in any given year and what the rules for calculating the deduction are.

[2] This is the same story as the above footnote. You'll have to find out what interest is taxable as income.

Table 3.6   Amortization Table for a $30,000 Loan with 8% Annual Interest, Compounded Monthly for 4 Years with $732.09 a Month Payments; "Simplified" Amortization Table

Payment #	Payment ($)	Balance ($)
0	0.00	35,154.72
1	732.39	34,422.33
2	732.39	33,689.94
3	732.39	32,957.55
4	732.39	32,225.16
5	732.39	31,492.77
6	732.39	30,760.38
7	732.39	30,027.99
8	732.39	29,295.60
9	732.39	28,563.21
10	732.39	27,830.82
11	732.39	27,098.43
12	732.39	26,366.04
13	732.39	25,633.65
14	732.39	24,901.26
15	732.39	24,168.87
16	732.39	23,436.48
17	732.39	22,704.09
18	732.39	21,971.70
19	732.39	21,239.31
20	732.39	20,506.92
21	732.39	19,774.53

In order to avoid having to explain why early payments on a loan go mostly to interest (and for some more insidious reasons that I'll get into in Chapter 5), some lenders multiply the monthly payment by the number of payments and call this the amount you owe (your balance) the day you take the loan. In the example of Table 3.4, this amounts to ($732.39)(48) = $35,154.72. They then present a table that looks like a true amortization table in which they simply take this very large number and subtract a payment each month, as is shown in Table 3.6. The arithmetic here is very simple. Since I got $35,154.72 by multiplying $732.39 by 48, when I subtract $732.39 from this "balance" 48 times, I get to 0, that is, the loan is paid off. As I have said, I'll discuss this further in Chapter 5. The biggest problem with this type of table is that it seems to be telling you that you owe $35,154.72 immediately after taking the loan. This is a lie (actually, it's just the first of many lies in this table). It is impossible for you to owe so much more than you borrowed almost immediately after you took the loan.

3.2 CALCULATING THE PAYMENT AMOUNT

Spreadsheets and online calculators handle this calculation very nicely. This section, deriving the formula, can be skipped if you wish.

If R is the annual interest rate and y is the number of payments per year, which we're assuming is the same as the number of compounding intervals per year, then $\dfrac{R}{y}$ is the interest per payment period. As a notational convenience, let

$$i = 1 + \frac{R}{y}.$$

The balance at the time of taking the loan, that is, after 0 payment periods, is just the principal. If we let B_n be the balance after the nth payment, then if P is the principal:

$$B_0 = P.$$

The balance after the first payment period is just the principal plus the interest accrued during this period minus the payment (S):

$$B_1 = P + P\frac{R}{y} - S = P\left(1 + \frac{R}{y}\right) - S = Pi - S.$$

This relationship is recursive. In other words, to get B_2, we use the same expression as above except that we replace B_0 with B_1,

$$B_2 = B_1 i - s = [Pi - S]i - s = Pi^2 - Si - S,$$

and then

$$B_3 = B_2 i - S = Pi^3 - Si^2 - Si - S.$$

The general expression is, therefore,

$$B_n = Pi^n - S\sum_{k=0}^{n-1} i^k.$$

This summation is actually a geometric series. If we let f be the summation,

$$f = \sum_{k=0}^{n-1} i^k = 1 + i + i^2 + \ldots + i^{n-1},$$

then

$$if = i + i^2 + i^3 + \ldots + i^n$$

and

$$if - f = f(i-1) = \left(i + i^2 + \ldots + i^{n-1} + i^n\right) - \left(1 + i + \ldots + i^{n-1}\right) = i^n - 1$$

and finally

$$f = \frac{i^n - 1}{i - 1}.$$

Putting this result back into the expression for B_n,

$$B_n = Pi^n - S\frac{i^n - 1}{i - 1}.$$

If we want the loan to be paid off at the end of n payment periods, then we want $B_n = 0$:

$$0 = Pi^n - S\frac{i^n - 1}{i - 1}.$$

Solving the above for S,

$$S = Pi^n \frac{i - 1}{i^n - 1},$$

and recalling the definition of i,

$$S = P\frac{R}{y}\frac{\left(1 + \frac{R}{y}\right)^n}{\left(1 + \frac{R}{y}\right)^n - 1}.$$

3.3 PAYING OFF A LOAN VERY SLOWLY

This section uses a little math, but I'll go through it slowly in small steps. As with the previous section, this section is not necessary if you don't want to tackle it.

I want to take a close look at what happens to the payment amount (S) as I take the same amount of loan, at the same interest rate, but for longer and longer times. In other words, I want to look at the payment formula above when I pick values for P, R, and y and leave them alone, but let n get larger and larger.

I'll introduce the term r for the interest per payment period, just to make things a little neater. That is, let

$$r = \frac{R}{y}$$

so that I can write the payment formula as the much neater looking

$$S = Pr\frac{(1 + r)^n}{(1 + r)^n - 1}.$$

r, the interest payment per period, is always a positive number. Therefore, when we add $1 + r$, we must always get a number greater than 1. n, the number of payments, is a positive integer such as 1, 2, 5, 20, or 240.

Since $(1 + r)^n$ just means $(1 + r)(1 + r)(1 + r) \ldots n$ times and $1 + r$ is bigger than 1, we can always be sure that $(1 + r) < (1 + r)^2 < (1 + r)^3$, and so on. In other words, as n (the number of payments to pay off the loan) gets larger, so does $(1 + r)^n$.

If n is very large, then $(1 + r)^n$ is much larger than 1, and $(1 + r)^n - 1 \approx (1 + r)^n$. This is a fancy way of saying that, for very large values of n, we can ignore the 1 in the denominator of the formula for S above.

This means that, for very large values of n,

$$S = Pr\frac{(1+r)^n}{(1+r)^n - 1} \approx Pr\frac{(1+r)^n}{(1+r)^n} = Pr.$$

But this simple value for S ($S = Pr$) is exactly the value of the first interest payment on the loan. Recall (or go back and look) that I have demonstrated that when your monthly payment is exactly the first interest payment, the balance never changes and you're just paying interest on the loan—forever. As n gets larger and larger and your payment approaches (slowly falls to) Pr, almost all of your payment goes toward the interest, and you are reducing your balance very, very slowly. This may or may not be a bad thing depending on your needs, your ability to pay, and other opportunities for the money you're not using to pay this loan. I'll go into some of these alternatives in later chapters.

3.4 MY WEBSITE SPREADSHEET

The spreadsheet Ch3Amortization.xls, Loan tab, produces an amortization table for a loan with regular monthly payments.

Table 3.7 is a snapshot of a part of this spreadsheet. This table distorts the spreadsheet slightly in that because the spreadsheet is too wide to fit on a printed page, I've taken the input data—the data to the left of the green line in the spreadsheet—and put it on top in the table.

The input variables for these calculations are

Start Month	the month when you take the loan; a number between 1 and 12;
Start Year	the year when you take the loan;
Nr Mnthly Pmts	if the loan is described in years, this is 12 (number of years);
Principal	the amount you're borrowing;
Rate	the annual percentage rate (APR).

The actual date of the loan is not shown.
The output columns are

Pmt Nr	a running count of the payments. Zero is the start date;
Mnth, Year	the month and year of each payment;
Balance	this is the balance *after* the current payment;
Payment	the monthly payment;
Interest	the interest accrued between last month and this month;
Tot Int/Year	a running total of the interest payments in each calendar year.

Table 3.7 Snapshot of the Loan Tab of Spreadsheet Ch3Amortization.xls

Start Month:	7
Start Year:	2005
Nr Mnthly Pmts	180
Principal	$350,000
Rate	6.94%

Pmt Nr	Mnth	Year	Balance ($)	Payment ($)	Interest ($)	Tot Int/Year ($)
0	7	2005	350,000.00	0.00	0.00	0.00
1	8	2005	348,890.00	3,134.17	2,024.17	2,024.17
2	9	2005	347,773.57	3,134.17	2,017.75	4,041.91
3	10	2005	346,650.69	3,134.17	2,011.29	6,053.20
4	11	2005	345,521.32	3,134.17	2,004.80	8,058.00
5	12	2005	344,385.42	3,134.17	1,998.26	10,056.27
6	1	2006	343,242.94	3,134.17	1,991.70	1,991.70
7	2	2006	342,093.86	3,134.17	1,985.09	3,976.78
8	3	2006	340,938.13	3,134.17	1,978.44	5,955.23
9	4	2006	339,775.72	3,134.17	1,971.76	7,926.99
10	5	2006	338,606.59	3,134.17	1,965.04	9,892.02
11	6	2006	337,430.69	3,134.17	1,958.27	11,850.30
12	7	2006	336,248.00	3,134.17	1,951.47	13,801.77
13	8	2006	335,058.46	3,134.17	1,944.63	15,746.41
14	9	2006	333,862.04	3,134.17	1,937.75	17,684.16
15	10	2006	332,658.71	3,134.17	1,930.84	19,615.00
16	11	2006	331,448.42	3,134.17	1,923.88	21,538.87
17	12	2006	330,231.12	3,134.17	1,916.88	23,455.75
18	1	2007	329,006.79	3,134.17	1,909.84	1,909.84
19	2	2007	327,775.38	3,134.17	1,902.76	3,812.59
20	3	2007	326,536.84	3,134.17	1,895.63	5,708.23
178	5	2020	6,214.38	3,134.17	53.75	444.89
179	6	2020	3,116.15	3,134.17	35.94	480.83
180	7	2020	0.00	3,134.17	18.02	498.85

3.5 A NOTE ABOUT TOTAL INTEREST FOR A YEAR

When adding up the total interest paid or received for a year for income tax purposes, it is important to look at the details very carefully. For example, suppose you are paying a loan back monthly, with payments due on the first day of the month. If you regularly send your payments right on time, your lender will be crediting your payment, approximately, on the eighth day of the month. An end of the year summary statement mailed to you for tax purposes in 2010 will show the 12 payments made in 2009.

Now, suppose that, for whatever reason, you mailed your January 2010 payment on December 20 of 2009. It is possible that it reached your lender and got credited to your account in 2009. This means that your 2009 statement will correctly show 13 credited payments. If you then go back to your regular schedule of mailing payments on the first of the month, your 2010 statement will show only 11 credited payments. There are no errors here; you just have to be careful to report what actually transpired.

My website spreadsheets are all set up assuming payments for a month are made or received on the first day of the following month. You can correct these calculations as needed for your situation.

3.6 ONLINE CALCULATORS

Again, there are so many of these available that I just picked a few:

1. www.bretwhissel.net/amortization/amortize.html;
2. www.bankrate.com/brm/popcalc2.asp;
3. www.vertex42.com/Calculators/loan-amortization-calculator.html.

This website provides a spreadsheet that you can download and use or modify: www.vertex42.com/ExcelTemplates/loan-amortization-schedule.html.

3.7 LOANS WITH FIRST PAYMENT DUE IMMEDIATELY

When you borrow cash, you don't usually make your first payment on the loan at the time you get your cash; instead, you wait for a month (or whatever the payment period is). When you never actually see the cash, for example, if you drive away in a financed new car, sometimes the first payment on the loan is due immediately rather than a month later.[3] Some online calculators let you take this variation into account.

The third tab in my spreadsheet, labeled Loan V2, calculates amortization tables for loans where the first payment is due at the date of the start of the loan.

In Table 3.8, I show (the first few lines of) a 15-year loan taken in May of 2008 for $175,000 at an APR of 8.00% with the first payment due immediately. As you can see, the format is the same as that of the Loan tab, so make sure you're using the correct tab when you study a loan.

---

[3] Even if you learn nothing else from this book, please memorize this paragraph: No two loan contracts are the same. In many cases, there are government regulations that guide the process, but there is still often quite a bit of room for *creativity* on the part of the lender. Never sign a loan contract until you have read and thoroughly understand both everything that you're agreeing to and all the implications of what you're agreeing to.

Table 3.8   A Loan With the First Payment Due When the Loan Is Taken

This sheet for loans with the first payment due at the start of the loan

Start Month:	5					
Start Year:	2008					
Nr Mnthly Pmts	180					
Principal	$175,000					
Rate	8.00%					

Pmt Nr	Mnth	Year	Balance ($)	Payment ($)	Interest ($)	Tot Int/Year ($)
1	5	2008	173,338.68	1,661.32	0.00	0.00
2	6	2008	172,832.96	1,661.32	1,155.59	1,155.59
3	7	2008	172,323.86	1,661.32	1,152.22	2,307.81
4	8	2008	171,811.37	1,661.32	1,148.83	3,456.64
5	9	2008	171,295.47	1,661.32	1,145.41	4,602.05
6	10	2008	170,776.12	1,661.32	1,141.97	5,744.02
7	11	2008	170,253.31	1,661.32	1,138.51	6,882.52
8	12	2008	169,727.02	1,661.32	1,135.02	8,017.55
9	1	2009	169,197.22	1,661.32	1,131.51	1,131.51
10	2	2009	168,663.88	1,661.32	1,127.98	2,259.49

The noticeable differences between the Loan V2 sheet and the earlier Loan sheet are that the Pmt Nr column starts with 1 rather than with 0. The first payment occurs at Pmt Nr 1, and the balance is reduced from the principal immediately. What's happening is that the borrower is receiving $175,000 and immediately returning $1,661.32. In other words, this is really a $173,338.68 loan.

3.8 IRREGULAR PAYMENTS

If you decide to make a payment that is larger than your regular payment, your balance will be reduced by an extra amount. Your lender should reduce your balance, and hence the accrued interest until your loan is paid off, accordingly. Similarly, if you make a payment that is smaller than your regular payment, some corrections must be made. In this latter case, penalties might be added to the balance as well as corrections to the calculations.

Different lenders handle these situations differently. This makes it impossible for me to discuss or present a spreadsheet that will handle the general case. Instead, I'll just show a couple of possibilities here. What I'm doing is mathematically correct, but remember that it might not apply directly to your loan.

I'll use the same loan as shown in Table 3.7. Suppose that in May 2006, I received a good income tax refund and I want to apply my newfound extra money to my loan. For payment number 10, I'll pay $10,000 instead of my regular $3,134.17.

Table 3.9 The Amortization Spreadsheet with Extra Payments Inserted

Pmt Nr	Mnth	Year	Balance ($)	Payment ($)	Interest ($)	Tot Int/Year ($)
0	7	2005	350,000.00	0.00	0.00	0.00
1	8	2005	348,890.00	3,134.17	2,024.17	2,024.17
2	9	2005	347,773.57	3,134.17	2,017.75	4,041.91
3	10	2005	346,650.69	3,134.17	2,011.29	6,053.20
4	11	2005	345,521.32	3,134.17	2,004.80	8,058.00
5	12	2005	344,385.42	3,134.17	1,998.26	10,056.27
6	1	2006	343,242.94	3,134.17	1,991.70	1,991.70
7	2	2006	342,093.86	3,134.17	1,985.09	3,976.78
8	3	2006	340,938.13	3,134.17	1,978.44	5,955.23
9	4	2006	339,775.72	3,134.17	1,971.76	7,926.99
10	5	2006	331,740.76	10,000.00	1,965.04	9,892.02
11	6	2006	330,588.71	3,070.62	1,918.57	11,810.59
12	7	2006	329,429.99	3,070.62	1,911.90	13,722.49
13	8	2006	328,264.57	3,070.62	1,905.20	15,627.70
14	9	2006	327,092.42	3,070.62	1,898.46	17,526.16
15	10	2006	326,984.10	2,000.00	1,891.68	19,417.85
16	11	2006	325,794.45	3,080.71	1,891.06	21,308.90
17	12	2006	324,597.93	3,080.71	1,884.18	23,193.08
18	1	2007	323,394.48	3,080.71	1,877.26	1,877.26
19	2	2007	322,184.07	3,080.71	1,870.30	3,747.56
20	3	2007	320,966.66	3,080.71	1,863.30	5,610.85

On my spreadsheet, just type $10,000 in for the correct payment as shown in Table 3.9. (If you haven't done it already, please take a minute and read the information in Chapter 15 on how to reestablish the original spreadsheet calculations after you've changed them.) As you can see, the payments after the $10,000 payment reflect the change and correctly show the new regular payment necessary to exactly pay off the loan at the end of 180 months. The balance, interest, and total interest per year columns are also correct.

Just to show what can be done, I changed payment 15 to a payment of $2,000.00. Again, the spreadsheet automatically corrected all of its calculations to reflect this change.

3.9 REGULAR SAVINGS

The other side of making regular payments to amortize a loan is making regular payments into a savings account so as to build up resources for your retirement, for a child's college education, and so on.

Suppose you could start saving $350 a month regularly. Savings bank interest rates will probably vary, but just to get an idea of what will happen, assume that they

Table 3.10   A Savings Plan Spreadsheet Example

Start Month:	7
Start Year:	2005
Nr Mnthly Pmts	360
Rate	4.50%
Monthly Pmt	$350.00

Pmt Nr	Mnth	Year	Balance ($)	Payment ($)	Interest ($)	Tot Int/Year ($)
1	7	2005	350.00	350.00	0.00	0.00
2	8	2005	701.31	350.00	1.31	1.31
3	9	2005	1,053.94	350.00	2.63	3.94
4	10	2005	1,407.89	350.00	3.95	7.89
5	11	2005	1,763.17	350.00	5.28	13.17
6	12	2005	2,119.79	350.00	6.61	19.79
7	1	2006	2,477.74	350.00	7.95	7.95
8	2	2006	2,837.03	350.00	9.29	17.24
9	3	2006	3,197.67	350.00	10.64	27.88
10	4	2006	3,559.66	350.00	11.99	39.87
11	5	2006	3,923.01	350.00	13.35	53.22
12	6	2006	4,287.72	350.00	14.71	67.93
184	10	2020	92,506.80	350.00	344.30	3,327.40
185	11	2020	93,203.70	350.00	346.90	3,674.30
186	12	2020	93,903.21	350.00	349.51	4,023.81
187	1	2021	94,605.35	350.00	352.14	352.14
356	2	2035	260,448.50	350.00	971.73	1,938.51
357	3	2035	261,775.18	350.00	976.68	2,915.19
358	4	2035	263,106.84	350.00	981.66	3,896.85
359	5	2035	264,443.49	350.00	986.65	4,883.50
360	6	2035	265,785.15	350.00	991.66	5,875.17

stay constant at 4.50% APR. (Another way to approach this is to save regularly for a year and then to use the year's savings to buy a higher interest certificate of deposit.)

Table 3.10 shows the results of using the Save tab on the same Ch3.Amortization. xls spreadsheet used above for loans. Again, I've put the items to the left of the green line on the spreadsheet on top, merely to fit on the printed page.

The input variables are similar to the loan input variables. The only difference is that the Principal variable on the Loan sheet has been replaced with a Monthly Pmt variable on the Save sheet.

The variables to the right of the green line are the same on both spreadsheets. In the Savings sheet, however, the balance and the interest grow each month. I've

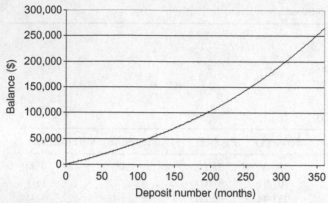

Figure 3.2 Example of growth of a regular savings plan.

included the total interest per year numbers in the Savings sheet because some savings' interest is taxable.

In the example shown, the monthly interest starts out very small. In the first full year of savings, the total interest for the year is not even $100. But look at what happens when you keep at it. At payment number 187 (about 15 years after starting), the monthly interest is greater than $350. In other words, the savings account itself is contributing more each month than you are. It's taken you 10 years to reach this point, but the rate of growth of your savings is getting large.

If you keep this up for 30 years as a savings plan toward your retirement, the monthly interest is over $900 and your balance is about $266,000. This is truly "the power of compound interest."

Figure 3.2 shows the balance versus the number of months that deposits have been made. As you can see, the instantaneous rate of growth is growing with time; that is, this savings account is "taking on a life of its own."

PROBLEMS

1. Find the regular monthly payments for the following loans (that accrue interest monthly):
   (a) Principal = $100,000; rate = 6.00%; 15 years
   (b) Principal = $230,000; rate = 9.10%; 20 years
   (c) Principal = $10,000; rate = 12.00%; 3 years
   (d) Without using a spreadsheet, principal = $250,000; rate = 6.00%; 15 years

2. Using the loan from problem 1a above, assume that the loan originated in July of 2005 and that the interest is deductible from your taxes. How much interest did you pay in the calendar years 2005, 2010, 2015, and 2020?

3. Using the loan from problem 2 above, after which payment (payment number, month, and year) did you pay off more than half of the principal?

4. Again using the loan from problem 2, in May of 2013, instead of your regular payment, you send in $10,000.00. Assuming that your lender is agreeable to recalculating, what will your regular payments be for the duration of the loan?

5. Again using the loan from problem 2, but start the loan in January for convenience, you have a job that pays you a small subsistence salary monthly and then sends you your commission checks quarterly. You negotiated a loan that allows you to pay interest only for 2 months and then to make a payment large enough to "catch up" on the third month. Show the first year's amortization table for this loan.

6. You have decided that you can make $1,000 a month payments on a 20-year loan. Calculate and plot a graph of the amount you can borrow versus the interest rate you can get, for interest rates varying from 0% to 10%. Remember that on a graph, you can't resolve too many significant figures, so it's all right to approximate the calculations.

   Many of the calculators on the Web will let you solve this problem directly—entering what you know and getting what you want. My spreadsheet is set up to calculate the payment from the other variables. To solve this problem on my spreadsheet, you'll have to enter the number of monthly payments and the rate, and then "adjust" the principal until the payment gets very close to $1,000. This problem illustrates the importance of getting as low an interest rate as you can to maximize your borrowing power for a fixed payment amount.

7. You are buying a new car for $31,800.00. You have $5,000 for a down payment and you wish to finance the remainder over 4 years. The car dealer offers you a 7% loan with up-front costs of $250. You take possession of the car on May 1, 2009 and your first loan payment is due the same day. How much are your monthly payments?

Chapter 4

Mortgages

The common use of the term *mortgage* today is that a mortgage is simply a loan you take when you buy a house. If I could leave it like that, this paragraph would be the entire chapter. After all, Chapter 3 discussed paying off loans, so what more is there to say?

Let's start by clearing up the definitions a bit. A mortgage is a common way of securing a loan with real property that you own. For most of us, real property simply means our house and the lot it sits on or our condominium. Typically, we take a loan to help pay for the house we are buying, and we secure the loan with a mortgage on the house. The terminology varies somewhat from state to state in the United States, but the idea is the same everywhere; you are pledging a property that you own as a security on your loan. The property is worth more than the amount of money you want to borrow, so that your lender doesn't have to worry about getting his or her money back.

Another common term is *second mortgage*. If you already have a loan for a part of the value of your house that is secured by a mortgage, and then you take a second loan that you secure with some of the remaining value of the house (some of your *equity* in your house), this is commonly known as a second mortgage.

Sometimes you'll take a *home equity loan*. A home equity loan is a loan that's secured with equity in your house, so it's very similar to a second mortgage. What's different is that a home equity loan often comes in the form of a checkbook that you can write checks on up to some predetermined limit. When you write a check on this account, you are actually borrowing some money, using a presigned contract for the terms of the loan. As you write checks and make payments, the balance of this loan goes up and down with the lender tracking everything and charging interest each month on the current balance.

Going back to mortgages, we don't have to stop at a second mortgage; there can be a third mortgage and so on. Unless you have significant real estate holdings such as an office building that's worth many millions of dollars, you might find that banks get a little nervous with something like the eighth mortgage holder, but conceptually, you can keep going as long as there's equity in your property. Lenders will

Understanding the Mathematics of Personal Finance: An Introduction to Financial Literacy, by Lawrence N. Dworsky
Copyright © 2009 John Wiley & Sons, Inc.

usually appraise the property and put a limit on how much they will loan, for example, up to 80% of the value of the property. This is because real estate values can change, and also, there are costs involved in a lender seizing a property if the borrower defaults on the loan.

In late 2008, real estate values dropped so drastically that many borrowers found that they owed more than their houses were worth. What happens in this unfortunate situation is not within the scope of this book. There are only two facts about this situation that I'm sure of: Neither borrowers nor lenders ever want to find themselves in this mess, and if it does happen, everybody is certain that somebody else's poor planning is at fault.

The bottom line is that what makes a mortgage a special kind of loan is that the lender has his or her money secured by your real property. For the typical borrower who is not planning to default on the loan, this is advantageous—a loan that gives the lender a superior, secured position over other existing secured creditors to the extent of repossessing the real estate is typically made available at lower interest rates than higher risk loans.

4.1 ONLINE CALCULATORS

These calculators are so numerous that I'm sure you can find many more in addition to the ones listed below. These look fairly good and are easy to use without first having to register or "get a quote." You'll also see some unusual numbers such as "average monthly interest." Please don't take these too seriously—read about Present Value in Chapter 7. Most of these sites will present a full amortization table if requested. This can be very convenient:

1. http://www.bankrate.com/brm/mortgage-calculator.asp;
2. http://www.homefair.com/tools/mortgage-payment-calculator/index.asp?cc=1;
3. http://realestate.yahoo.com/calculators/payment.html.

4.2 FIXED RATE MORTGAGES

Mortgage loan interest rates can be arranged in any of several ways. First, there is the fixed interest loan. This is what I have presented in the last chapter. The interest rate stays the same for the entire repayment period. Typically, you commit to pay the loan back in a fixed period such as monthly payments for 15 years and fixed payments are calculated. In most cases, you can make extra payments (or increase the amount of some of your payments). This, of course, reduces your outstanding balance and therefore reduces the interest charged. I'll go through some examples of these variations below. In some cases, you're allowed to pay just the interest, in which case the balance stays the same. Most mortgages do not come with prepayment penalties. This means that there is no charge for paying off or reducing the

Table 4.1   Fixed Rate, Fixed Payment Mortgage Amortization Table

Pmt Nr	Month	Year	Balance ($)	Payment ($)	Interest ($)	Tot Int/Year ($)
0	7	2009	200,000.00	0.00	0.00	0.00
1	8	2009	199,567.14	1,432.86	1,000.00	1,000.00
2	9	2009	199,132.11	1,432.86	997.84	1,997.84
3	10	2009	198,694.91	1,432.86	995.66	2,993.50
4	11	2009	198,255.52	1,432.86	993.47	3,986.97
5	12	2009	197,813.94	1,432.86	991.28	4,978.25
6	1	2010	197,370.15	1,432.86	989.07	989.07
7	2	2010	196,924.13	1,432.86	986.85	1,975.92
8	3	2010	196,475.89	1,432.86	984.62	2,960.54
9	4	2010	196,025.41	1,432.86	982.38	3,942.92
238	5	2029	2,844.37	1,432.86	21.28	176.28
239	6	2029	1,425.73	1,432.86	14.22	190.50
240	7	2029	0.00	1,432.86	7.13	197.63

balance on the loan sooner than the original schedule prescribes. Don't assume too much here—this is an important point and you should read your mortgage loan contract very carefully before you sign it.

For analyzing a basic fixed interest, fixed payment loan, the website references given in Chapter 3 will do the job. If you're going to vary the payments, a simple calculator can't adequately do the job; a spreadsheet is definitely the way to go. I'll start with a simple spreadsheet and fixed payments just so that I can review how things are set up. Then, I'll show how to modify this to accept any payments whatsoever and to update all information and predictions. The spreadsheet Ch4Mortgages.xls on my website handles all of the calculations in this chapter.

The first tab on my spreadsheet, Basic, is identical to the Loan tab on the Chapter 3 spreadsheet. The use is the same, and the examples are the same. The only difference is that now we call the loan a mortgage loan.

As an example, consider a $200,000 principal and a 20-year (240 monthly payment) mortgage loan at 6.00% annual percentage rate (APR). Table 4.1 shows the spreadsheet results. I did not include the input region (to the left of the green line) because this is identical to the examples in the previous chapter. The loan is taken on a day in July 2009.

Looking at Table 4.1, or the spreadsheet, the regular monthly payment is $1,432.86.

If your loan agreement allows it, you might decide to just make interest payments or a reduced amount payment for some number of months. Assume that at payment number 15, in October 2010, you want to make interest-only payments for a few months. The interest accrued in the month coming up to payment number 15 is $968.69. If you change the payment amount at this time to just this amount, you'll see that the interest for payment 16 changes to this same amount. This

Table 4.2 Repeat of Table 4.1 with Interest-Only and a Single Large Payment Inserted

Pmt Nr	Month	Year	Balance ($)	Payment ($)	Interest ($)	Tot Int/Year($)
12	7	2010	194,660.40	1,432.86	975.59	6,876.50
13	8	2010	194,200.84	1,432.86	973.30	7,849.80
14	9	2010	193,738.98	1,432.86	971.00	8,820.81
15	10	2010	193,738.99	968.69	968.69	9,789.50
16	11	2010	193,738.99	968.69	968.69	10,758.20
17	12	2010	193,739.00	968.69	968.69	11,726.89
18	1	2011	193,739.00	968.69	968.69	968.69
19	2	2011	193,739.01	968.69	968.70	1,937.39
20	3	2011	193,739.01	968.69	968.70	2,906.09
21	4	2011	193,253.68	1,454.03	968.70	3,874.78
22	5	2011	192,765.92	1,454.03	966.27	4,841.05
23	6	2011	192,275.73	1,454.03	963.83	5,804.88
24	7	2011	191,783.08	1,454.03	961.38	6,766.26
25	8	2011	187,741.99	5,000.00	958.92	7,725.17
26	9	2011	187,253.63	1,427.07	938.71	8,663.88
27	10	2011	186,762.82	1,427.07	936.27	9,600.15
28	11	2011	186,269.57	1,427.07	933.81	10,533.96
29	12	2011	185,773.84	1,427.07	931.35	11,465.31

is correct—since you haven't changed the balance at payment 15, the interest accrued the following month should be the same as the interest accrued the previous month.

If you haven't done it already, please take a minute and read the information in Chapter 15 on how to reestablish the original spreadsheet calculations after you've changed them.

In Table 4.2, you'll see the same loan as shown in Table 4.1 but with interest-only payments being made for 6 months. At payment number 21, the regular monthly payment has jumped from $1,432.86 to $1,454.03 to pay off the loan at the end of 240 months (from the date of taking the loan).

Also shown in Table 4.2 is the result of making a $5,000 payment at payment number 25. After this payment, the regular monthly payment drops to $1,427.07.

4.3 ADJUSTABLE RATE MORTGAGES (ARMs)

Not all mortgage loans commit to a certain rate of interest for the life of the loan. The ones that don't commit to this are called, reasonably enough, ARMs.

An ARM will typically offer an attractive low rate for a certain period of time (e.g., 5 or 10 years) after which time the rate can change. Just how much it can change, how often it can change, and what factors will be used to calculate a change will be specified in the contract. Also, there are laws that limit just how aggressively the lender can change the interest rate. Typically, the new mortgage rates will be tied to various federal government interest rates.

An ARM can be a very attractive way to get into a new home with modest monthly payments. The initial ARM rates are typically lower than fixed interest mortgage loan rates because the lender does not have to worry as much about how interest rates might vary in the future and therefore does not have to conservatively "cover his or her bets." .

On the other hand, if you are just managing to make your mortgage loan payments and then your interest rate jumps, say, 1% on a $300,000 loan balance, you suddenly could be looking at approximately $250 per month higher payments. Add this to the fact that your real estate taxes have almost definitely gone up every year and you might find yourself unable to make your payments.

The ARM tab in my spreadsheet lets you examine various changing rate scenarios. Comparing the ARM tab to the Basic tab, you'll see that the rate entry in the Basic tab has been replaced by a Rate column in the ARM tab. If you enter the initial Rate at the top of this column (cell H2), this rate is immediately propagated down the column. In Table 4.3, I repeated the previous example (with the original payment schedule restored). Five years into the loan (at payment number 60), the mortgage rate changes to 7% and then at payment number 72 it changes to 8%. These numbers are put in by first entering 7% in the Rate column at payment number 60 and then entering 8% in the Rate column at payment number 72. The new rates propagate

Table 4.3   ARM Amortization Table

Pmt Nr	Month	Year	Rate (%)	Balance ($)	Payment ($)	Interest ($)	Tot Int/ Year($)
57	4	2014	6.00	171,533.42	1,432.86	860.53	3,459.14
58	5	2014	6.00	170,958.23	1,432.86	857.67	4,316.81
59	6	2014	6.00	170,380.16	1,432.86	854.79	5,171.60
60	7	2014	7.00	169,799.20	1,432.86	851.90	6,023.50
61	8	2014	7.00	169,263.49	1,526.20	990.50	7,014.00
62	9	2014	7.00	168,724.66	1,526.20	987.37	8,001.37
63	10	2014	7.00	168,182.68	1,526.20	984.23	8,985.60
64	11	2014	7.00	167,637.54	1,526.20	981.07	9,966.66
65	12	2014	7.00	167,089.23	1,526.20	977.89	10,944.55
66	1	2015	7.00	166,537.71	1,526.20	974.69	974.69
67	2	2015	7.00	165,982.98	1,526.20	971.47	1,946.16
68	3	2015	7.00	165,425.01	1,526.20	968.23	2,914.39
69	4	2015	7.00	164,863.78	1,526.20	964.98	3,879.37
70	5	2015	7.00	164,299.28	1,526.20	961.71	4,841.08
71	6	2015	7.00	163,731.49	1,526.20	958.41	5,799.49
72	7	2015	8.00	163,160.39	1,526.20	955.10	6,754.59
73	8	2015	8.00	162,630.69	1,617.44	1,087.74	7,842.32
74	9	2015	8.00	162,097.45	1,617.44	1,084.20	8,926.53

down the Rate column automatically. In Table 4.3, or directly on the spreadsheet, you see that the original $1,432.86 monthly payment jumps to $1,526.20 at the first-rate jump and then to $1,617.44 at the second-rate jump.

Some ARM mortgages have the annoying attribute that any payment larger than the original monthly payment might trigger a change in the APR. In other words, the original initial guaranteed fixed rate period of the loan ends immediately.

4.4 BALLOON LOANS

A balloon loan is one in which the term of the loan (i.e., the number of payments) is just a number on which the monthly payments is calculated. After a relatively small number of years (5 or 10), the loan suddenly "balloons," and the entire balance is then due in one lump payment. Assuming that you don't happen to have this money in the bank, you can suddenly find yourself scrambling for a new loan with which to pay off the old loan. If this happens to you, hope that interest rates have gone down, or at least haven't gone up very much since you took the loan.

In my spreadsheet, on either tab, the balloon payment at a given payment number is just the balance plus the monthly payment (remember that the balance entry is the outstanding loan balance after the monthly payment has been made).

Looking at Table 4.3, for example, the loan payoff amount at payment number 72 is $163,160.39 + $1,526.20 = $164,686.59.

4.5 UP-FRONT COSTS

You won't see the term up-front costs in your mortgage contract. I'm using the term as a catchall for all the costs and fees involved in starting up a mortgage loan. This includes points. Points are a start-up fee that many lenders charge for giving you the loan. One point represents 1% of the loan amount. Then there are appraisal fees, paperwork fees, various state and county taxes, and so on. While you as the borrower certainly want to scrutinize every one of these items and make sure you're getting the best deal available (i.e., the lowest amount of up-front money necessary to get yourself the loan), for my purposes, I'm just going to sum them all into up-front costs.

Many lenders will offer you several deals, for example, "6% plus 3.5 points, or 7% plus 2 points." It is therefore not obvious which is the best deal, and it's neces-sary to do some calculations before making your choice.

Some costs are tax deductible and therefore the loan actually costs you less than it appears when you take the loan. Without knowing your tax bracket, I can't take this "discount" into account. If your tax bracket[1] is, say, 30%, then you can estimate that you'll save almost 30% on the deductible costs.

Suppose you want all the costs folded into the loan. Assume that there are $10,000 in up-front costs. I'll go back to my first example, the one that's loaded onto

---

[1] Tax brackets are discussed in Chapter 9. Very quickly for now, it's the tax rate you're paying on the top few dollars of your income and is not a bad estimate of how much you'll get back from the IRS when you report deductible expenses.

the "Basic" tab on the spreadsheet we've been using. Just to recap, in case you've changed the data and don't want to bother reloading the spreadsheet from my website, you are taking a $200,000 loan at 6.00% interest, to be paid back (amortized) over 240 monthly payments (20 years). The monthly payment is $1,432.86.

If you want your lender to fold the $10,000 that you need to come up with into the loan itself, then, from the lender's point of view, you're getting a $210,000 loan.

Using the same spreadsheet, if you change the $200,000 principal to $210,000, the regular monthly payment goes up to $1,468.68 (or, by scaling, just note that $210,000 is 5% greater than $200,000 and multiply $1,432.86 by 1.05). Write this number down, or just enter it into a vacant cell on the spreadsheet so that you can refer to it.

Change the principal in the spreadsheet back to $200,000. The regular payment drops back to $1,432.86. Start increasing the interest rate by small amounts until the payment equals to $1,468.68. At an interest rate of approximately 6.31%, this goal is accomplished. This is the effective interest rate on your loan of $200,000 when the $10,000 extra is absorbed by the lender.

PROBLEMS

Fixed interest, fixed payment mortgage loans are no different from the basic loans described in Chapter 3. The problems below will, therefore, concentrate on variations in calculation brought about by different types of mortgage loans. Solving these problems, on the other hand, will require a working ability to solve the simpler problems.

For problems 1–4, start with a basic mortgage loan of $325,000 taken in January 2010, payable monthly for 20 years with an APR of 5.80%.

1. Your up-front costs for getting your loan were 3 points and $450.00 in fees. You didn't have this money available, so you had these costs folded into the loan. What's your effective interest rate?

2. Since you are short on cash, you make interest-only payments on the above loan for the first 3 years. How much are these payments, what is your balance at the end of 3 years, and what is your new regular payment to amortize the loan?

3. Continuing with the above situation, after 5 years (Pmt Nr 60), interest rates drop and you are offered a free refinancing of the loan at 5.00%. You'd like to pull some cash out for other purposes and you're "used to" paying $2,577.66 each month. Your new mortgage period is again 20 years, but now it's 20 years from the start of your new mortgage. How much is the new mortgage for and how much cash do you pull out?

4. I bought a house that was appraised at $425,000 on January 1, 2000. I got an 80% mortgage as a fixed 30-year mortgage at 5.00%. On January 1, 2007, I decided to take a second mortgage because I need some cash. I find that my house has appreciated 3% a year. My lender will give me a second mortgage for the difference between 80% of the appraised value and my equity. The second mortgage is at 6.2%. The second mortgage will be fully paid off at the same time that the first mortgage is paid off.

What is the principal of the second mortgage and what are my total monthly payments (first + second mortgages combined)? Assume 0 up-front costs for both mortgages.

Chapter 5

Prepayment Penalties

When you take a loan, the lender has legitimate up-front costs in preparing the paperwork, setting up the account, monitoring the payments, and so on. In the case of a mortgage, as shown in Chapter 4, the lender usually manages to charge you for these costs. In the case of an auto loan, there might not be any up-front costs. Instead, the lender estimates his or her costs and wraps them into the interest rate for the loan. This is equivalent to what I have shown at the end of Chapter 4; it's just not shown explicitly—you're quoted an interest rate and that's that.

Suppose you were to acquire some money you didn't think you'd have, or you have the savings, or for whatever reason you decide to pay the loan off early. The lender not only loses expected interest, but he or she also loses the part of the up-front costs that haven't been recaptured yet. In this case, it's not unreasonable for the lender to charge you these unrecaptured costs in the form of a *prepayment penalty*.

How a prepayment penalty is calculated must be spelled out in the original loan agreement. Unfortunately, historically, some lenders have looked on this situation as an opportunity to grab some extra profits. In many states, laws have been enacted to limit the prepayment penalty that lenders can demand.

One way of keeping you from realizing just how much the prepayment penalty is costing you is to keep you from realizing just how much your outstanding balance is. Table 5.1 shows a simple 3-year car loan. The principal is $15,000 and the interest rate is 8.0%. The monthly payment is $470.05. As you can see by looking at the balance column, the balance correctly goes to 0 after 36 payments.

If you add up all the interest payments, you get $1,921.64. Now add this to the principal and you get $16,921.64. This is of course the same number you get if you add up all the payments (or equivalently, multiply one payment amount by the number of payments). This number is shown at the top of the right-hand column, which I've labeled *fake balance*. If you deduct the 36 payments from this amount, one per month, you again get to a 0 balance at the end of 36 months.

If you are presented with the *fake balance* column as a payment schedule instead of the real *balance* column, you are not being cheated in that you are paying the correct monthly payment and paying off the loan at the correct time. But, along the way, you have no idea what your real outstanding balance is.

Table 5.1 Auto Loan with "Interest Up-Front" Balance

Pmt Nr	Balance ($)	Interest ($)	Fake balance ($)
0	15,000.00	0.00	16,921.64
1	14,629.95	100.00	16,451.59
2	14,257.44	97.53	15,981.55
3	13,882.45	95.05	15,511.50
4	13,504.95	92.55	15,041.46
5	13,124.94	90.03	14,571.41
6	12,742.39	87.50	14,101.36
7	12,357.30	84.95	13,631.32
8	11,969.63	82.38	13,161.27
9	11,579.38	79.80	12,691.23
10	11,186.53	77.20	12,221.18
11	10,791.07	74.58	11,751.14
12	10,392.96	71.94	11,281.09
13	9,992.20	69.29	10,811.05
14	9,588.77	66.61	10,341.00
15	9,182.65	63.93	9,870.96
16	8,773.82	61.22	9,400.91
17	8,362.27	58.49	8,930.86
18	7,947.97	55.75	8,460.82
19	7,530.91	52.99	7,990.77
20	7,111.07	50.21	7,520.73
21	6,688.44	47.41	7,050.68
22	6,262.98	44.59	6,580.64
23	5,834.69	41.75	6,110.59
24	5,403.54	38.90	5,640.55
25	4,969.52	36.02	5,170.50
26	4,532.60	33.13	4,700.45
27	4,092.77	30.22	4,230.41
28	3,650.01	27.29	3,760.36
29	3,204.30	24.33	3,290.32
30	2,755.62	21.36	2,820.27
31	2,303.94	18.37	2,350.23
32	1,849.26	15.36	1,880.18
33	1,391.54	12.33	1,410.14
34	930.77	9.28	940.09
35	466.93	6.21	470.05
36	0.00	3.11	0.00

For example, just after your twelfth payment, 1 year after you took the loan, your outstanding balance is $10,392.96. With no prepayment penalties, this is what it should cost you to pay off the loan. Typically, you'd make the twelfth payment and the outstanding balance payment at the same time. Also, lenders are usually prorating interest by the day, so depending on which day you actually want to pay off the loan, your exact payment should be close to but not exactly this number.

Now, look at the fake balance for the same day—$11,281.09. This is almost $1,000 more than the actual balance, about 8.5% more. If the lender quotes a number such as $10,700 to pay off the loan after your twelfth payment and your only information is the fake balance payment schedule, you can't see that you're being charged about $300 prepayment penalty. You'll see some information such as "unearned interest" on the payoff document as an explanation of why your payoff number is lower than the fake balance.

This latter type of calculation is sometimes called a *precomputed* loan calculation.

5.1 RULE OF 78

This rule is sometimes called the sum-of-digits method. I'll explain where both of these names came from shortly. As you shall see, any loan bearing this type of prepayment penalty is to be avoided. Various states have a legislation limiting its use. Hopefully, by the time you're reading this, the practice will have been banned universally.

The rule of 78 payoff calculation used to be very popular with automobile loans. I understand it is uncommon today. As long as it remains legal, however, you'll see it being used—principally among somewhat unscrupulous dealers who offer "subprime" loans to people who don't have good credit, probably at an exorbitant interest rate.

I'll use the auto loan of Table 5.1 as my first example. The spreadsheet Ch5PrepaymentPenalties.xls is used to calculate these examples. The data input area to the left of the green line is identical to the data input area in the previous chapters' spreadsheets, so I needn't repeat it in detail. Also, the first six columns to the right of the green line are identical to the same columns in the previous chapters' spreadsheets. Since there are so many numbers to calculate, this spreadsheet is quite wide and you'll have to scroll a bit to see everything. If you like, you can shrink the width or hide the columns carrying intermediate calculates (such as shares) to see the results easily.

The first number I need is the sum of the digits of the payment numbers: $1 + 2 + 3 + \ldots + 35 + 36 = 666$. If you were to do this for a 1-year loan, you'd get $1 + 2 + 3 + \ldots + 11 + 12 = 78$, hence the name (although I'm pretty sure there aren't many 1-year auto or other similar loans).

Next, assuming the loan was to be fully paid off, I need the sum of all the interest components of the monthly payments. In Table 5.1, this is the sum of all the numbers in the column labeled *interest*, which is equal to $1,921.64.

Divide the sum of the interest payments by the sum of the payment numbers and call the result the interest fraction:

$$\text{Interest fraction} = \frac{1,921.64}{666} = 2.885.$$

So far, all I've done is to calculate an odd parameter, the interest fraction. Now I'm going to multiply this interest fraction by the payment number, but in reverse order, and call this the number of shares.

For payment number 1, I'll multiply the interest fraction by 36, from which I get $(2.885)(36) = 103.87$. For payment number 2, I'll have $(2.885)(35) = 100.99$.

Table 5.2 is a repeat of Table 5.1 but with the *fake balance* column removed and several new columns, including a shares column, added.

This shares calculation is where the injustice comes about. If you add up all the entries in the shares column, you get $1,921.64. Similarly, if you add up all the entries in the interest column (new column #4), you get the same number. However, look at the total interest (Tot Int) column—this is a running total of the interest payments—and the "earned interest" column—a running total of the shares column. The name earned interest was coined by the people who came up with this procedure. It doesn't mean what it says. The shares column loads the interest payments up front, early in the loan, so that "earned interest" isn't really earned interest accrued; it's a conveniently faked number.

The payoff number is calculated by adding together the true balance and the difference between the earned interest and the actual total interest. For example, after payment number 12,

$$\text{Payoff} = \$10,392.96 + \$1,056.03 - \$1,033.51 = \$10,415.49.$$

Compare this with the actual payoff at the same time, which is $10,392.96. The actual prepayment penalty is the difference between the payoff and the balance, which, after payment number 12, is $22.53.

Figure 5.1 shows the prepayment penalty for this loan versus the payment number. As the figure shows, 1 year into the loan, the prepayment penalty peaks at about $22.

At this point, you're probably wondering if all of this was worth the trouble; $22 is not a huge amount of money, and I did say that the lender is often entitled to some prepayment penalty.

Now consider a $300,000, 15-year loan at the same interest rate (8%). Figure 5.2 shows the prepayment penalty as a function of payment number for this loan. About 5 years into the loan, this penalty is almost $12,000! When you consider that the administrative and bookkeeping costs of this loan aren't much different from that of the previous example, this is one very, very large prepayment penalty.

While I've provided an online worksheet for calculating rule of 78 prepayment penalties, the rule of thumb is that you don't want to be dealing with a lender who asks for this in the loan agreement. If it's a small several-year loan such as the example in Table 5.2, then you can probably live with it because the amount isn't very large. In general, however, it's close to robbery.

Table 5.2 Auto Loan with Rule of 78 Prepayment Penalty Added

Pmt Nr	Balance ($)	Interest ($)	Tot Int ($)	Shares	Earned interest ($)	Payoff ($)	Penalty ($)
0	15,000.00	0.00	0.00				
1	14,629.95	100.00	100.00	103.87	103.87	14,633.83	3.87
2	14,257.44	97.53	197.53	100.99	204.86	14,264.77	7.33
3	13,882.45	95.05	292.58	98.10	302.96	13,892.82	10.38
4	13,504.95	92.55	385.13	95.22	398.18	13,518.00	13.04
5	13,124.94	90.03	475.17	92.33	490.51	13,140.28	15.34
6	12,742.39	87.50	562.66	89.45	579.95	12,759.68	17.29
7	12,357.30	84.95	647.61	86.56	666.51	12,376.20	18.90
8	11,969.63	82.38	730.00	83.67	750.19	11,989.82	20.19
9	11,579.38	79.80	809.79	80.79	830.98	11,600.57	21.18
10	11,186.53	77.20	886.99	77.90	908.88	11,208.43	21.89
11	10,791.07	74.58	961.57	75.02	983.90	10,813.40	22.33
12	10,392.96	71.94	1,033.51	72.13	1,056.03	10,415.49	22.53
13	9,992.20	69.29	1,102.79	69.25	1,125.28	10,014.69	22.49
14	9,588.77	66.61	1,169.41	66.36	1,191.65	9,611.01	22.24
15	9,182.65	63.93	1,233.33	63.48	1,255.12	9,204.44	21.79
16	8,773.82	61.22	1,294.55	60.59	1,315.72	8,794.99	21.16
17	8,362.27	58.49	1,353.04	57.71	1,373.42	8,382.65	20.38
18	7,947.97	55.75	1,408.79	54.82	1,428.24	7,967.43	19.45
19	7,530.91	52.99	1,461.78	51.94	1,480.18	7,549.32	18.40
20	7,111.07	50.21	1,511.98	49.05	1,529.23	7,128.32	17.25
21	6,688.44	47.41	1,559.39	46.17	1,575.40	6,704.44	16.01
22	6,262.98	44.59	1,603.98	43.28	1,618.68	6,277.68	14.70
23	5,834.69	41.75	1,645.73	40.39	1,659.07	5,848.03	13.34
24	5,403.54	38.90	1,684.63	37.51	1,696.58	5,415.49	11.95
25	4,969.52	36.02	1,720.66	34.62	1,731.20	4,980.07	10.55
26	4,532.60	33.13	1,753.79	31.74	1,762.94	4,541.76	9.16
27	4,092.77	30.22	1,784.00	28.85	1,791.80	4,100.57	7.79
28	3,650.01	27.29	1,811.29	25.97	1,817.77	3,656.49	6.48
29	3,204.30	24.33	1,835.62	23.08	1,840.85	3,209.53	5.23
30	2,755.62	21.36	1,856.98	20.20	1,861.05	2,759.68	4.06
31	2,303.94	18.37	1,875.35	17.31	1,878.36	2,306.95	3.00
32	1,849.26	15.36	1,890.71	14.43	1,892.78	1,851.33	2.07
33	1,391.54	12.33	1,903.04	11.54	1,904.33	1,392.82	1.28
34	930.77	9.28	1,912.32	8.66	1,912.98	931.43	0.66
35	466.93	6.21	1,918.52	5.77	1,918.75	467.16	0.23
36	0.00	3.11	1,921.64	2.89	1,921.64	0.00	0.00

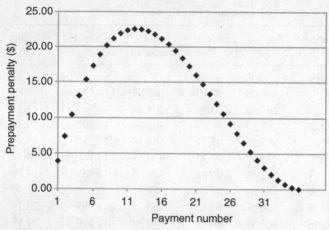

Figure 5.1   Prepayment penalty versus payment number with data from Table 5.2.

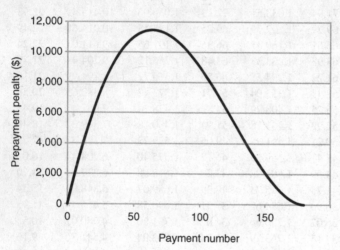

Figure 5.2   Prepayment penalty versus payment number for a large loan example.

5.2   OTHER PREPAYMENT PENALTIES

The list of how many different forms of prepayment penalties can be contrived is endless. Here are a few examples I have encountered:

1. Six months' interest on 80% of the balance. Looking at the rule of 78 example shown in Figure 5.2, after 5 years, the remaining balance is approximately $236,000. Eighty percent of this is approximately $189,000, and 6 months' interest on this is about $7,560. This is considerably better than the rule of 78 prepayment penalty but still a lot of money.

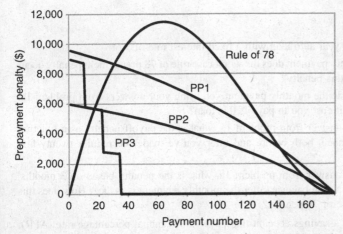

Figure 5.3 Comparison of different penalties for various prepayment penalty formulas (large loan example).

2. A flat 2% of the outstanding balance. In the above example, this would be about $3,600. This is still a lot, but we're getting better.

3. Three percent of the balance for the first year of the loan, 2% for the second year, 1% for the third year, then 0 for the remainder of the loan term. This is very expensive for the first year of the loan ($9,000) but obviously not of concern after the third year.

4. Innumerable variations on the above. That is, a declining penalty that goes to 0 after some reasonable (i.e., short) period of time. The lender's goal here is to keep you from jumping ship if rates drop before the lender has recovered at least a reasonable piece of the up-front expenses.

Figure 5.3 shows the rule of 78 prepayment penalty and also the first three of the above prepayment penalties (labeled PP1, PP2, and PP3). Interestingly enough, very early in the loan, the rule of 78 penalty is the best of the group (best = lowest penalty). For most of the loan, this is approximately from 36 to 145 months; the rule of 78 penalty is the worst of the group.

Many loans, particularly mortgages, allow for a fairly substantial partial prepayment, with the accompanying drop in interest amount, without triggering a prepayment penalty (if there is one). Also, many state laws have something to say about prepayment penalties on mortgages. Another point: Some prepayment penalties are triggered either by the sale of the home or the prepayment of the mortgage without the sale, and some are triggered only by the prepayment of the mortgage. You'll have to do some research to find out what applies to you. Please do all of these the day before you sign the loan contract, not the day after. Also, remember that you're the customer; when you take a loan—negotiate.

PROBLEMS

1. Consider a $25,000, 5-year auto loan with a fixed interest rate of 7.6%.

   (a) After which monthly payment does the worst case rule of 78 prepayment penalty occur and how much is this penalty?

   (b) If you haven't made the monthly payment yet, using your answers from problem 1a, how much would it cost you to pay off this loan?

   (c) Again using your answer from problem 1a, what is the payoff penalty as a function of the unpaid balance, both before and after you've made the regular twenty-first payment?

   (d) Again using your answer from problem 1a, what is the penalty based on 3 months' interest on the remaining balance (after the monthly payment is made)? How does this compare to the rule of 78 penalty?

2. Assume that you have a savings account that pays 7.00% annual percentage rate (APR), compounded monthly, with $100,000 in it on the day you bought your car. The best money-managing philosophy would have been to take the money from your savings account and to buy your car with cash because you are paying more on your loan than your money was earning.[1] For whatever reason, you took the loan and are making your payments by withdrawing money from your savings. You now decide that you made a mistake and want to pay off the loan. In both the prepayment penalty examples shown above and also the case of no prepayment penalty, when is the best time to pay off your loan?

3. Repeat the same problem but with a savings interest rate of 8.0%.

---

[1] I'm ignoring subtle issues such as inflation and taxation in this problem.

Chapter 6

Credit Cards

Credit cards are such an integral part of our society that it's hard to imagine a time when they weren't around. Store and gasoline credit cards have a long history, but the popularity of bank credit cards dates back only to the late 1960s. Today, it's easy to get a card—either a bank, a store, or a gasoline card. Maybe it's a little bit too easy to get a card; many people have several of them. It's easy to make purchases; you just present the card. Electronic card readers collect your information and communicate with your credit card company almost instantly. Sometimes, it's not so easy to fully pay for the purchases. And it's incredibly easy to accrue balances on several different credit cards—balances that never seem to go away.

Gasoline and store credit cards are intended for purchases at the card issuers' gasoline station or store. Bank credit cards on the other hand are intended for use just about anywhere.[1] Most stores and gasoline stations will accept them for purchases even if they also issue their own cards. Bank credit cards also offer instant loans, which are called cash advances.

Tracking the calculation of your credit card interest (also called the *finance charge*) is not as easy as it is with a mortgage or an auto loan. You make purchases and possibly take cash advances many times each month, at random times and for differing amounts. Every month, a statement comes from the credit card company detailing all of these transactions. The monthly statement also shows a date for paying for all of your purchases without incurring any interest (the *due* date). If you always religiously pay your full balance on or before the due date, you do not pay any interest on your credit card purchases.

Credit cards can be a pretty good deal. Many credit cards are free. When you make a purchase using a credit card, you are actually borrowing money from a bank. You do not pay purchase interest charges if you regularly pay your bill in full before the due date shown on the statement. Interest-free loans are hard to come by. The obvious question that comes to mind then is, "Who is paying for this loan?" The answer is that the vendor of your purchase is paying for the loan. The vendor pays a fee for each purchase, usually calculated as a percentage of the amount of the purchase.

---

[1] One of the first of these cards, issued sometime around 1969, was actually called The Everything Card.

Understanding the Mathematics of Personal Finance: An Introduction to Financial Literacy, by Lawrence N. Dworsky
Copyright © 2009 John Wiley & Sons, Inc.

Vendors' fees are not your problem except, of course, that these fees drive up the prices of your purchases. In a sense, this is fair. You're paying for the convenience of being able to charge your purchases. The economics and marketing aspects of this belong in a business studies book, not this book. Our topic of interest (pun fully intended) is what happens when you can't, or won't, fully pay your bill each month. You then have to pay the interest. For the rest of this chapter, I'll use the terms *interest* and *finance charge* interchangeably. Credit card companies seem to prefer the latter term, but I prefer the former term because it emphasizes the parallels and differences between credit card interest and other forms of debt interest.

Not fully paying your balance each month is alluring. The credit card statement shows a "minimum payment" that will keep the credit card company happy for a month. This minimum payment might be much less than the total amount of the bill. Typically, we're all a little short of money; we'd like to go on with our lives (and continue making our purchases) as comfortably as possible, so we stare longingly at the minimum payment number and compare it to the number for fully paying for the month's transactions (plus any remaining balance from previous months' unpaid or incompletely paid transactions).

6.1   CREDIT CARD STATEMENTS

Each credit card company's monthly credit card statement has a unique layout, so I can't specifically explain your statement to you. Statements are similar, however, in having the following sections:

- a tear-off piece showing your name and address, the card company's name and address, all or part of your account number, your account balance, the payment due date, and possibly the minimum payment due. This piece is returned to the card company with your payment;
- an account summary showing your previous balance, payments and credits during the previous month, finance charges for the previous month, and your new balance;
- a transaction summary detailing your purchases, cash advances, and payments for the month;
- a finance charge summary showing the different rates charged and your specific finance charges for the month;
- a list of all transactions with the dates of these transactions and usually a very terse explanation to help identify the transaction;
- an information section (usually on the back of the sheet) explaining how to report a disputed item, how your finance charges are calculated, how the minimum payment is calculated, and so on.

Other information may also be found in any or all of these sections. Also, a several page document comes with the credit card itself—this is your contract with the credit card company. It includes many more details of how your account will be

Amazing
Card

Account balance: $139.36

Minimum payment: $15.00

Cardholder: Harry Consumer
123 Sunny Lane
Scottsdale, AZ 85258

Enter amount enclosed:

$

Acct # 1234 5678 9876 5432

Payment due date: 04/20/09

Closing date: 03/26/09

Credit line: $12,000.00

PO Box 12345
New York, NY 10034-4321

Cash or credit available: $11,840.64

Figure 6.1 Fictional credit card payment sheet.

Finance charge schedule

Category	Daily periodic rate	NAPR
Balance transfers	0.016164%	5.90%
Purchases	0.016164%	5.90%
Cash advances	0.032328%	11.80%

Figure 6.2 Fictional credit charge schedule.

handled. This contract is revised occasionally; the revisions come on a separate piece of paper that accompanies your statement (or might possibly come in the mail by itself).

In Figures 6.1–6.3, I've created some fictitious generic credit card statement sections. They're not exactly sections from any statement that I know of, but they should be close enough to your statement sections for you to be able to find and understand the relevant information.

Figure 6.1 shows the tear-off piece of the statement. In this example, the balance isn't very high—only $139.36. The due date for the payment is April 20, 2009. The minimum payment that must be made by this date is $15.00. If at least $15.00 doesn't

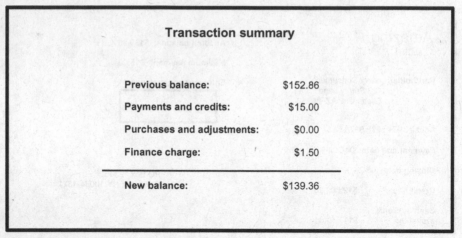

Transaction summary

Previous balance:	$152.86
Payments and credits:	$15.00
Purchases and adjustments:	$0.00
Finance charge:	$1.50
New balance:	$139.36

Figure 6.3   Fictional credit card transaction summary sheet.

reach the Amazing Card bank by April 20, 2009, late charges will be applied in addition to the interest accrued. If $139.36 reaches the bank by April 20, 2009, there might not be any new interest charges—depending on factors that will be discussed below.

What if I pay at least $15.00 but less than $139.36 before the due date? How much interest must I pay on an unpaid balance? How is the unpaid balance calculated? To find the answers to these questions, I must now look to the other sheets (or sections of sheets) on my statement.

Figure 6.2 shows the finance charge schedule for this credit card. I'll talk about balance transfers in a bit. The interest rate for purchases is 5.90% annual percentage rate (APR). The interest rate for cash advances is higher—13.80% APR. In both cases, the daily periodic rate is just the nominal annual percentage rate (NAPR) divided by 365. The difference between the two rates will be relevant when I discuss how payments are allocated. Remember that these are fictitious numbers. In some cases, the cash advance and purchase rates are the same; in some cases, they're higher or lower than shown here.

Figure 6.3 shows the account (activity) summary for the month.[2] Last month's balance had been $152.86. A $15.00 payment had been made, leaving a balance of about $137.86. The word "about" is needed here because the actual date of the payment is important, as will be shown soon. If I assume that a full 31-day month has gone by since the last payment and that this is a purchase balance, the interest due should be ($137.86)(0.00016164)(31) = about 70 cents. Why is the listed finance charge $1.50?

The answer to this lies in the details found either on the back of the statement or in the card contract. There is a minimum finance charge of $1.50. In a sense,

---

[2] I should actually use the term "billing cycle" rather than month here. Although the period is 1-month long, it did not start at the beginning of the month.

the credit card company is charging you extra for not owing them as much money as they'd like you to owe them. This finance charge is more than 13% APR.

The transaction list should be scrutinized for errors and possible fraudulent entries. Compare listed purchases to your receipts. Do not delay in reporting discrepancies.

The back of the statement presents the details of how your credit card company calculates your balance and finance charge and how it credits your payments. Since money is coming and going (purchase, cash advances, payments) pretty much on a daily basis, it makes sense for the credit card lenders to calculate interest on a daily basis. This means that we need to understand how the daily balance is calculated.

First, however, I'll discuss balance transfers.

6.2 TRANSFERS

Transfers are "deals" offered by a credit card company to lure you away from another credit card company. Each offered deal is unique, so you might have to do a little work to evaluate, compare, and contrast them.

Let me create a fictional transfer deal: If you transfer your balance from credit card company X to credit card company Y, company Y will allow you to maintain that balance, and possibly some level of new purchases, for 12 months (from the transfer date) at 0% interest. This can be a very good offer. Remember that there is always a time value of money, so an offer to assume a debt of yours (your outstanding balance at company X) for a year with no interest is in effect paying you to switch credit card companies.

Just how much this deal is worth depends on what your balance with company X is, how much interest you're paying company X, and what your intentions and abilities to pay off this debt are.

The spreadsheet Ch6CardBalanceTransfer.xls on my website lets you analyze various scenarios. This spreadsheet compares two alternatives, that is, keeping your original card with taking a transfer offer. The spreadsheet assumes that you have the savings balance from which to make payments. It then compares your net worth at the end of some number of payment periods for the two alternatives. In other words, what I'm doing here is calculating the present value of the two alternatives.

In creating this spreadsheet, I assumed that you will make regular payments on time. Missing or late payments invoke penalty fees as well as extra interest, and unless you know your upcoming late/missing payment schedule, there's no way to calculate any of this beforehand. The spreadsheet assumes that you have an amount twice your card balance (an arbitrary choice) in a bank account on day 1 and that all payments are made out of this bank account.

Your net worth in this example is the difference between your savings account balance and your credit card balance.

Table 6.1 Example of a Credit Card Balance Transfer Deal

			Keep original credit card			Switch cards		
		Pmt Nr	Bank balance ($)	Card X balance ($)	Net worth ($)	Bank balance ($)	Card Y balance ($)	Net worth ($)
		0	24,000	12,000	12,000	24,000	12,000	12,000
Nr Pmts	12	1	23,880	11,980	11,900	23,880	11,860	12,020
Pmt	$200	2	23,760	11,960	11,800	23,760	11,719	12,040
Balance	$12,000	3	23,639	11,939	11,700	23,639	11,578	12,061
Rate of old card	18.00%	4	23,518	11,918	11,599	23,518	11,436	12,082
Rate of new card	6.00%	5	23,396	11,897	11,499	23,396	11,293	12,103
Rate of savings	4.00%	6	23,274	11,875	11,399	23,274	11,149	12,125
		7	23,152	11,854	11,298	23,152	11,005	12,146
		8	23,029	11,831	11,197	23,029	10,860	12,169
		9	22,905	11,809	11,097	22,905	10,715	12,191
		10	22,782	11,786	10,996	22,782	10,568	12,214
		11	22,658	11,763	10,895	22,658	10,421	12,237
		12	22,533	11,739	10,794	22,533	10,273	12,260

Consider the following example: You are offered a 1-year (12-month) transfer deal that lets you move your existing balance of $12,000 from an existing 18% APR account (Card X) to a 6% APR account (Card Y). Your savings are earning 4%.[3] The monthly payment you plan to make is $250.

Table 6.1 shows this spreadsheet. Your net worth at the end of the year jumps from $10,835 to $12,266 if you switch cards. Since the purpose of this spreadsheet is to make comparisons rather than to calculate a detailed balance, I'm showing amounts only to the nearest dollar.

If you get an offer of 0% interest for switching, your net worth jumps to $12,922 at the end of the year (enter 0% in the rate new card cell). You can change any or all of the variables to the left of the green line arbitrarily and reversibly, evaluating every transfer offer you receive.

A few words to the wise about transfers. First of all, if you take one of these deals and then spend the money that would have gone for payments on something else, you come out in a mess at the end of the year. If this money is needed for a family health or other emergency, well, you just do it—but otherwise this is a poor strategy.

Next, why on earth would a credit card company want to give you an interest-free or very low-rate loan for a year (or 9 months, or for however long it offers

[3] In this situation, all financial logic says to take the money from the savings account and pay off your credit card balance. My purpose here is to compare transferring your balance to not transferring your balance, so I'll pass on this choice.

the deal)? Clearly, in the grand scheme of things, it must be making more money by offering these deals than by not offering them. There are four answers to this puzzle:

1. Card Y has lured you away from card X. Card Y has gained a customer and card X has lost a customer. Remember that credit card companies also make money from the vendors when you charge a purchase, and you will be making more purchases (and possibly paying more interest) in the future.

2. Many people lose track of the upcoming due date when the deal ends. If the balance, or the agreed upon part of the balance, is not paid by this upcoming due date, various fees and/or back interest payments will be charged.

3. Many people divert payment money to buy other stuff. Often, they'll spend even more than the payment money they've diverted. This means that at the end of the year, they've not only spent all the money that would have gone to the original balance, but they've accrued even more new credit card debt.

4. Read your credit card paperwork carefully. Certain balance transfer offers may take away the grace period on purchases. I'll discuss more about the importance of the grace period below, but in a nutshell, this is saying that in some cases, even if you pay for new purchases the day the statement arrives in your mailbox, you'll still be paying interest on these new purchases to the credit card company.[4]

6.3 PAYMENT ALLOCATION

When you are only partially paying your balances (and paying interest), you will be paying toward two separate balances: your purchase totals and your cash advance totals (I'm leaving possible existing transfer situations out right now.) How much of each payment goes toward the purchase balance and how much of it goes toward the cash advance balance?

Payments are usually allocated to first pay off the balance that carries the lowest interest rate. This is typically the purchase balance. As an example, if your cash advance balance is $500 and your purchase balance is $125, a $100 payment will go entirely toward reducing your purchase balance to $25. This allocation is totally in the interest of credit card companies (again, pun fully intended). They are maximizing their profits.

---

[4] On the other hand, there are some transfer offers that combine 0% interest on transfers and 0% interest on purchases for a year. Since there are some things that you must buy, for example, food for you and your family, this is essentially an offer of extra money for you—if you handle it responsibly.

6.4 DAILY BALANCE

The daily balances of your purchases and/or cash advances are the amounts on which finance charges are calculated. To understand the details of the finance charge calculation, you first need to understand the calculation of the daily balances for each of the days in the billing period (usually a month). Note that there are two separate balances: one for purchases and one for cash advances.

To calculate a day's daily balances,

1. start with the previous day's daily balances;
2. add transactions and fees charged that day (to the appropriate balance);
3. add finance charges accrued on previous day's daily balance when appropriate;
4. subtract any credits and payments applied that day against the balances. Remember to apply as much of the payment as possible to the lower interest balance (typically the purchase balance).

6.5 SOME CALCULATION EXAMPLES

For the following examples, I'm going to start each month (which will have 30 days and will also be the billing period) with the same initial situation:

1. The daily interest numbers are 0.03011% for purchases and 0.06299% for cash advances.
2. At the end of the previous month, my purchase daily balance was $335.25, and my cash advance daily balance was $480.20.
3. For the previous month, my purchase average daily balance (ADB) was $253.77, and my cash advance ADB was $435.90. Don't worry yet about how to calculate the ADB—you'll see it done several times below.
4. The interest is calculated as the (ADB)(daily interest rate)(number of days in the month). For my purchases, this was

$$(\$253.77)(0.03011\%)(30) = \$2.29,$$

and for my cash advances, it was

$$(\$176.50)(0.06299)(30) = \$3.34.$$

Adding the interests to the end of the month daily balances, I got a total purchase balance of $337.54 and a total cash advance balance of $483.54, adding up to a total balance of $821.08. These are the numbers that appear on my statement at the end of the month.

I received my statement a few days later. I made my payment and it was received and credited by the bank on the twelfth day of the following month, well before the statement due date.

EXAMPLE 6.1   *Table 6.2*

I pay my bill in full ($821.08) and vow to never use my credit card again. As you can see from the table, my balances just come forward day after day until day 12. Since I have sent in the full payment, everything gets zeroed out and my daily balances are 0 from the twelfth day until the end of the month.

At the end of the month, the bank again calculates the ADB. This is a simple calculation—in each category, add up all the daily balances and then divide this number by the number of days in the month. Shown in the table are the purchase ADB of $123.77 and the cash advance ADB of $177.29.

These ADBs in turn create interest payments of $1.12 and $3.35 for the purchase and cash advance ADBs, which in turn add up to a total new balance of $4.47. This amount is billed. It takes 2 months of paying bills fully to clear my account. This will be relevant when I discuss grace periods below.

EXAMPLE 6.2   *Table 6.3*

This example is the same as the previous example except that I didn't have the cash on hand to fully pay my bill. Instead, I paid $300.

The credit card company first applies a payment to the lowest interest-bearing balance in the account, in this case, the purchase balance. Since the purchase balance of $337.54 is higher than the payment of $300, the purchase balance absorbs the entire payment.

For the rest of the month, the daily purchase balance is $37.54, and the cash advance balance is $483.54.

At the end of the month, the ADBs are calculated as before, as are the month's finance charges. I end the month owing $38.89 on my purchase balance and $492.98 on my cash advance balance. My monthly statement shows a total due of $531.86.

If these balances were being carried on two different credit cards, I could have optimized things a bit. Rather than paying $300 to the purchase balance, I would have paid the minimum payment required to the purchase balance, and then paid the remainder of my $300 to the cash advance balance. Since the cash advance balance accrues such a high interest rate in this example, the annoyance of writing two checks and the cost of the extra postage stamp would have been well worth the effort. I should of course watch out for incurring a minimum finance charge.

EXAMPLE 6.3   *Table 6.4*

This example extends the previous examples to show a fairly busy month. There are three purchases, two cash advances, and one payment. In this example, I made the payment larger than the purchase daily balance, so the purchase daily balance is fully paid off and some of the payment gets applied to the cash advance daily balance.

Notice the trend as time progresses for the cash advance balance, the balance that accrues interest at a very high rate. This slowly becomes the dominant part of the total balance.

I didn't create an online spreadsheet to do the work shown in these tables. Your credit card company does this for you and sends it to you every month.

The very busy interweaving of dates on which money comes into and goes out of your account every day is what necessitates the system of daily tallying of charges and payments. It's complicated but not unnecessarily complicated.

Table 6.2   Example 6.1: Hypothetical Monthly Statement and Calculation Detail

Day	Item	Purchases	Advances	Credits ($)	Purchase daily balance ($)	Advance daily balance ($)
1					337.54	483.54
2					337.54	483.54
3					337.54	483.54
4					337.54	483.54
5					337.54	483.54
6					337.54	483.54
7					337.54	483.54
8					337.54	483.54
9					337.54	483.54
10					337.54	483.54
11					337.54	483.54
12	Pmt			821.08	0.00	0.00
13					0.00	0.00
14					0.00	0.00
15					0.00	0.00
15					0.00	0.00
16					0.00	0.00
17					0.00	0.00
18					0.00	0.00
19					0.00	0.00
20					0.00	0.00
21					0.00	0.00
22					0.00	0.00
23					0.00	0.00
24					0.00	0.00
25					0.00	0.00
26					0.00	0.00
27					0.00	0.00
28					0.00	0.00
29					0.00	0.00
30					0.00	0.00
				ADB:	123.77	177.29
				Interest:	1.12	3.35
				Balance due:	1.12	3.35
		Total due:	$4.47			

Table 6.3   Example 6.2: Hypothetical Monthly Statement and Calculation Detail

Day	Item	Purchases	Advances	Credits ($)	Purchase daily balance ($)	Advance daily balance ($)
1					337.54	483.54
2					337.54	483.54
3					337.54	483.54
4					337.54	483.54
5					337.54	483.54
6					337.54	483.54
7					337.54	483.54
8					337.54	483.54
9					337.54	483.54
10					337.54	483.54
11					337.54	483.54
12	Pmt			300.00	37.54	483.54
13					37.54	483.54
14					37.54	483.54
15					37.54	483.54
15					37.54	483.54
16					37.54	483.54
17					37.54	483.54
18					37.54	483.54
19					37.54	483.54
20					37.54	483.54
21					37.54	483.54
22					37.54	483.54
23					37.54	483.54
24					37.54	483.54
25					37.54	483.54
26					37.54	483.54
27					37.54	483.54
28					37.54	483.54
29					37.54	483.54
30					37.54	483.54
				ADB:	148.79	499.65
				Interest:	1.34	9.44
				Balance due:	38.89	492.98
		Total due:	$531.86			

Table 6.4   Example 6.3: Hypothetical Monthly Statement and Calculation Detail

Day	Item	Purchases ($)	Advances ($)	Credits ($)	Purchase daily balance ($)	Advance daily balance ($)
1					337.54	483.54
2					337.54	483.54
3					337.54	483.54
4					337.54	483.54
5	Purchase	150.00			487.54	483.54
6					487.54	483.54
7					487.54	483.54
8	Advance		100.00		487.54	583.54
9					487.54	583.54
10					487.54	583.54
11					487.54	583.54
12	Pmt			500.00	0.00	571.08
13					0.00	571.08
14					0.00	571.08
15					0.00	571.08
15	Advance		100.00		0.00	671.08
16					0.00	671.08
17					0.00	671.08
18					0.00	671.08
19					0.00	671.08
20	Purchase	75.00			75.00	671.08
21					75.00	671.08
22					75.00	671.08
23					75.00	671.08
24	Purchase	5.80			80.80	671.08
25					80.80	671.08
26					80.80	671.08
27					80.80	671.08
28					80.80	671.08
29					80.80	671.08
30					80.80	671.08
				ADB:	187.62	624.68
				Interest:	1.69	11.80
				Balance due:	82.50	682.88
			Total due:	$765.38		

6.6   GRACE PERIOD

The grace period is the period of time during which you are offered 0% interest on your purchases if you pay off your total balance in full. Let me break out a few of these points carefully. For the bank cards that I'm familiar with, there is no grace period on cash advances. You start accumulating interest when you get the cash and the accounting is done as shown in Tables 6.2–6.4. To get a grace period on your purchases, you must pay off the total new balance, not just your purchase balance, on your statement within the grace period.

How long is the grace period? Typically, almost a month. However, if you're getting your statement in the mail (as opposed to online), then 5–10 days can go by after the closing date before you see the statement. Also, you need to allocate a few days for your check to get to the credit card company. If you get your statement online and pay your bills online, you'll have almost an entire month of "grace."

How do you fall in or out of grace? You must have paid your previous bill in full and on time to be qualified for a grace period on your current bill, and, of course, you must pay the current bill in full and on time.

6.7   CHANGING INTEREST RATES

If a credit card company decides that your risk as a debtor has changed, the credit card company might increase your interest rates immediately. This decision can be triggered by many factors in your financial life and need not be due to any history of late or insufficient payments to this credit card company.

One situation, so egregious that lawmakers have addressed it, is that some credit card agreements allow the credit card companies to change interest rates not only on new purchases and cash advances but also on existing balances. This is analogous to, for example, the bank that gave you an auto loan calling you sometime and telling you to "throw away your payment schedule, we've raised your interest rate and we'll be sending you a new payment schedule with higher payments."

As I keep repeating, I'm in no position to give legal advice—not only because I'm not a lawyer but also because some banking laws are federal laws, some are state laws that vary from state to state, and all legislation is subject to change. Read your credit agreement when you get a new card, and read the occasional "your agreement has changed" notices. If you agreed to the contract (and using the card is considered accepting the agreement), you don't have much of a leg to stand on when you realize that you don't like it. You can cancel the card, but this won't affect balances and interest rates already in place.

6.8   A BANKRUPTCY SPIRAL

Table 6.5 shows a month-by-month account of how things can go very, very, wrong. In doing the calculations, I'm just looking at monthly updates—I'll assume that you

Table 6.5   A Credit Card Disaster

		Month	Balance ($)	Interest ($)	Tot Int ($)
Pmt	$200.00	0	254.50	2.00	2.00
Charge	$450.00	1	511.55	7.05	9.05
		2	771.16	5.12	
		3	1,033.37	7.71	
		4	1,298.21	10.33	
		5	1,565.69	12.98	
		6	1,835.84	15.66	
		27	8,176.86	78.44	
		28	8,513.12	81.77	
		29	8,852.75	85.13	
		30	9,195.78	88.53	
		31	9,542.24	91.96	
		32	9,892.16	95.42	
		33	10,245.58	98.92	
		34	10,602.54	102.46	
		35	10,963.07	106.03	
		36	11,327.20	109.63	

pay your bill at the end of the billing period and make your purchases on the first day of the billing period.

Suppose your new credit card purchase interest rate is 1% per month. In your budget, you see that you can pay $200 a month toward your credit card bill. Unfortunately, the lifestyle you have chosen requires $450 a month in purchases.

At first, things don't look so bad. You're only paying a few dollars a month in interest, and look at the purchases, restaurant meals, and vacation trips you're getting. What's so terrible?

What is so terrible is what's creeping up month after month and is about to explode. Look at payment 33: At that time you're paying almost $100 each month in interest.[5] Half of your payment isn't buying you anything. Your balance has exceeded $10,000. That's a lot of debt. Your card is probably "maxed out," that is, you have the maximum balance that your credit card company will allow, and your ability to charge new purchases is cut off.

Now you have a useless credit card and you're $10,000 in debt. What are you going to do? If you're really financially suicidal, you'll transfer the balance to a new credit card with a higher debt limit and for a short while, once again, go on your merry way.

---

[5] I'm being a little sloppy here. Actually, as your balance approaches $10,000, the minimum payment might get larger than $200. This is a small correction and doesn't affect the point(s) that I want to make.

6.9 MINIMUM PAYMENT

Every credit card statement shows a minimum payment or a minimum amount due. This is the least amount of money that you can send in (before the due date) that will not trigger late payment fees or fines. Typically, this amount is sufficient to reduce your balance a bit—if you haven't recently made more payments or received more cash advances.

The calculation of the minimum payment varies from credit card company to credit card company and can be slightly complicated. Usually there's a lowest minimum payment (something like $15). If your balance due is less than this amount, then your minimum payment will be your balance due. If you don't want to work through the details of the minimum payment for your credit card, remember that once your balance is significantly higher than the lowest minimum payment, the minimum payment will be enough to reduce your balance by a little bit (assuming that there are no new transactions) but not by much.

6.10 OTHER INTEREST CALCULATION APPROACHES

The ADB method of calculating finance charges that I presented above is not the only way that finance charges are calculated. Other approaches include the adjusted balance, the two-cycle ADB, the previous balance, and the ending balance. As best as I could ascertain, the ADB method is used by major credit card companies in the United States.

If you wish to learn more about the other approaches, try these websites:

1. www.finweb.com/banking-credit/how-credit-card-finance-charges-are-calculated.html;

2. http://money.howstuffworks.com/personal-finance/debt-management/credit-card8.htm;

3. http://www.bankrate.com/brm/green/cc/basics3-2a.asp.

6.11 DEBIT CARDS—SOMETHING COMPLETELY DIFFERENT

A debit card looks almost exactly like a credit card. It is, however, completely different from a credit card. When you charge a purchase or get a cash advance with a credit card, you are borrowing money. When you pay for a purchase or get cash with a debit card, you are accessing your own money from a designated bank account. It's pretty much the same as writing a check except that it's more convenient, and the electronic fund transfers occur immediately. Also, many stores are wary of taking a check because the check might bounce (your account did not have enough money in it) or the check might be fraudulent or forged to start with. When you use a debit card, the store verifies the available funds in your account and gets its money immediately.

There are two common ways to use a debit card. You can pay for, say, groceries at the supermarket with your debit card. When you swipe your card and then enter your secret personal identification number (PIN) into the little machine, your purchase is debited from your bank account. Typically, at the same time, the machine asks if you'd like some cash back. If you say yes and then enter the amount you'd like, the store cashier will hand you cash along with your receipt—as if you had paid cash and were getting change. This cash is of course deducted from your account along with your purchase. There are usually no fees for purchases or purchases accompanied by cash back.

The second way of using a debit card is to go to an automatic teller machine (ATM). These machines let you swipe your card, enter your PIN, and request a cash disbursement. These disbursements are offered in discrete amounts such as $60, $80, $100, and so on. Because the machine is only loaded with certain denomination bills (in this case $20 bills), an arbitrary amount request cannot be handled. Often, ATMs will charge a fee for the transaction. This fee can be a combination of a fee from your bank and a fee from the owner of the ATM if it is not your bank's ATM. Read everything you can find at an ATM—on the front of the machine, on the wall above the machine, on the screen as you enter information, and of course on your debit card to make sure that you understand what fees will be charged. These fees are not interest because there is no time element involved.

Remember that cash advances accrue interest starting from the day you take the cash advance, with a minimum interest (finance charge) applied. This means that the percentage cost of taking a cash advance varies directly with the amount of money you take and the length of time you take to repay this money (the minimum finance charge could skew this calculation a bit). On the other hand, the percentage cost of cash taken from an ATM not owned by your bank is high for small advances and low for large advances. Also, since it's your own money that you're using rather than borrowed money, you lose savings or checking account interest until you repay the money to your account.

PROBLEMS

1. Assume a purchase interest rate of 0.0333% per day. The current month has 30 days, and the billing period is from day 1 to day 30. You have no previous balances.

   (a) You charge a $150 purchase on the first day of the month. What do you owe if you pay this fully before the due date?

   (b) Your credit card company receives your payment of $100 on the third day of the next billing period. This payment is received before the due date so there are no penalties. What is your daily balance on the second day of the next billing period and on the third day of the next billing period?

   (c) Assuming that this "next" billing period has 31 days, what is your purchase balance at the end of this month's statement (assuming that you don't use the card at all)?

2. You are always a little short of cash before your payday, which is the last day of each month. The last day of your credit card billing period is also the last day of each month.

On the twentieth day of each month, you get a cash advance of $250. Since you see your credit card bills electronically, your paycheck is automatically deposited into your checking account and you pay your bills from your checking account electronically; there are no unanticipated delays. On the first day of every month, you pay your credit card bill, and it reaches the credit card company on the third day of the month. Assume a cash advance interest rate of 0.0500% per day and that every month has 31 days. How much is your monthly credit card bill for your cash advances? Assume that you fully pay your bill each month.

3. Assume a purchase daily interest rate of 0.03% and a cash advance interest rate of 0.06%. You make a purchase of $150 on the first day of the month and take a cash advance of $250 on the twentieth day of the month (a 31-day month). You will pay your bill on the first day of the following month, and it will be received by the credit card company 2 days later (on the third day of that month). You do not have enough money to pay the entire bill when you first receive it, so you pay 50% of the bill. Assuming that you never use your card again and after the first payment you fully pay your bills, calculate the first two bills and payments.

Chapter 7

Present Value

Suppose you told me that you had

- walked into a television store,
- handed the store owner $900, and
- walked out with a television that sold for $1,000.

Assuming your story was true, I would conclude either that you were an amazing negotiator or that you were dealing with a very unintelligent (and soon to be out of business) store owner.

But what if your story was true (i.e., all the facts presented were absolutely correct), but something was omitted? Let's retell the story with an omitted step: You

- walked into a television store,
- handed the store owner $900,
- came back 2 years later, and
- walked out with a television that sold for $1,000.

This is no longer an interesting story. If the store owner had taken your money, put it in a savings bank account at (approximately) 5% interest and waited 2 years, he or she would have $1,000 and would certainly be willing to hand over a $1,000 television set.

The more common way of doing this is for you to put your $900 in your bank for 2 years, withdraw your total $1,000, and go buy your $1,000 television. Financially, there is no difference between the two approaches. Note that this is *not* buying on credit. I've discussed this in another chapter.

This simple story exemplifies a basic concept of financial dealings: Nothing makes sense unless you include the effects of time when comparing costs, prices, bank account values, and so on. In the story above, you are loaning the store owner $900 for 2 years and applying the interest on this loan, along with your original $900, to purchase the television.

Let me present a slightly different example. Suppose today is January 1. You want to buy a used car and take possession of it today. Unfortunately, you don't have

Understanding the Mathematics of Personal Finance: An Introduction to Financial Literacy, by
Lawrence N. Dworsky
Copyright © 2009 John Wiley & Sons, Inc.

enough money to pay fully today, so you agree to pay $1,000 today and $1,000 on each upcoming January 1 for the next 4 years (a total of five 1,000 payments). What did this car cost you?

The simple answer, $5,000, is incorrect. In the television purchase story above, didn't you buy a $1,000 television 2 years from now for only $900 in your hand today? Using the same logic, doesn't the $1,000 payment 2 years from today only cost you approximately $900 today?

The repeated reference to *today* in the above paragraph is the key to the concept of present value. You're taking possession of the car today. The present value of the car is what it would cost if you could pay cash for it today. Since you will be paying for the car in five payments, each at a different time in the future, you need to figure out what each of these payments is worth today. (Remember, the $1,000 payment 2 years from today is only worth approximately $900 today.) The correct way to do this is to figure out how much money you would have to deposit into a savings bank on January 1 so that if you withdrew $1,000 from the savings bank every subsequent January 1, the savings account would have the exact amount of money to make these payments—there would be no money left after the fifth payment, and the account wouldn't run out of money before making the fifth payment.

This is an absolutely fair way of doing things. Once you've calculated this present value you could put it in the bank and make the annual payments, or you could just give this present value to the car seller, and he or she in turn could put it in the bank and extract $1,000 every year for 5 years. By agreeing to sell the car for this payment plan, the car seller has effectively agreed that the sum of the present values of all the payments is exactly today's selling price of the car.

I'll go through this present value calculation slowly here just to show what's happening, then I'll show how to do it a bit more efficiently on a spreadsheet. Finally, I'll show some online calculators that make it all very simple.

Assume that my savings bank is paying 5% interest, compounded annually. Annual compounding is not very typical, but it lets me go through the calculations without burying you in pages of numbers. On the spreadsheet to follow, I'll use a more typical monthly compounding.

Start with the first payment: $1,000 today is worth $1,000 today. This is a very trivial thing to say, but it's worth saying to help point out the contrast to future payments.

Next, look at the second payment: $1,000 a year from today. If I put $1 in the bank today, a year from today I would have $1 + 5% of $1 = (1.05) × $1 = $1.05. Going the other way, if I want to have $1,000 a year from today, I need to deposit $1,000/1.05 = $952.38.

For the third payment, I need to look at two compounding intervals. I've already looked at the result of one compounding interval, that is, $1,000 2 years from today is worth $952.38 1 year from today, which in turn is worth $952.38/1.05 = $907.03 today (this is the number that I approximated as $900 for the television set example).

Table 7.1 shows the present value of each of the five payments. The longer you can wait before making a payment, the less that payment costs you. Since each of

Table 7.1 Car Purchase Present Value Calculations

Payment number	Years from now	Present value per payment ($)
1	0	1,000.00
2	1	952.38
3	2	907.83
4	3	863.84
5	4	922.70
	Total present value:	4,545.95

the five numbers in Table 7.1 is a present value, it makes sense to add them up and to calculate your actual purchase price for the car, which is $4,545.95.

The above calculation also demonstrates what actually happens when you buy a car on time. If the purchase price of the car is $4,546 and you agree to buy it "on time" with annual payments at an annual percentage rate (APR) of 5% (compounded annually), then your annual payment would be $1,000.

7.1 ONLINE CALCULATORS

With all calculators, but especially with present value calculators, read the accompanying materials very carefully. In many cases, these calculators are giving you the present value of some amounts of money in a complicated situation. This situation might or might not be what you are looking for:

1. www.moneychimp.com/calculator/present_value_calculator.htm;
2. www.uic.edu/classes/actg/actg500/pfvatutor.htm;
3. www.investopedia.com/calculator/PVCal.aspx.

7.2 DOING IT WITH MY SPREADSHEET

The spreadsheet Ch7PresentValue on my website performs present value calculations. Table 7.2 shows the present value of a 12 monthly payment example. The payments are each $250 and the APR is 5.00%. In this situation, you're making the first payment on the first day of the calculation. The twelfth monthly payment therefore occurs 11 months after day 1.

On the right of Table 7.2 you see the Pmt Nr column and then the Payment column. The spreadsheet is initialized with all the payments being equal. The next column is the Payment PV column. This column shows the present value of each payment on day 1. As you can see, the further out in time a payment is made, the lower its present value. This is because the present value represents the amount of money that you would have to put into a savings account at the stated interest rate, to have the payment number at the time you are actually making the payment.

Table 7.2   Spreadsheet Present Value Example

	Pmt Nr	Payment ($)	Payment PV ($)	Balance ($)
		Total PV = $2,920.31		
Nr Pmts: 12	1	250.00	250.00	250.00
Payments: $250	2	250.00	248.96	501.04
	3	250.00	247.93	753.13
Rate: 5.00%	4	250.00	246.90	1,006.27
	5	250.00	245.88	1,260.46
	6	250.00	244.86	1,515.71
	7	250.00	243.84	1,772.03
	8	250.00	242.83	2,029.41
	9	250.00	241.82	2,287.87
	10	250.00	240.82	2,547.40
	11	250.00	239.82	2,808.01
	12	250.00	238.82	3,069.71

The last column in the table (and on the spreadsheet) is the hypothetical account balance. This is the balance you would find in your checking account at the time of making each payment (immediately after you made the payment).

The total present value of all the payments is at the top of the table (and the spreadsheet). This is simply the sum of all the payment present values.

Just to check that things are making sense, if you deposited $2,932.47 on day 0 (and then left the account alone), at the end of 11 months your balance would have grown to

$$\$2,932.47(1+0.05/12)^{11} = \$3,069.71,$$

which is indeed your balance at the end of 11 months.

As with all of my spreadsheets, you can change the numbers to the left of the green line as much and as often as you like. If you change any entries to the right of the green line, you must pay attention to resetting the original spreadsheet when necessary (see Chapter 15 on how to do this).

This spreadsheet can handle more complicated situations than the one described above. For example, if you want Pmt Nr 6 to be 0 (you missed a month) but for Pmt Nr 7 you put in $500, just type these entries into the appropriate cells in the spreadsheet. The Total PV will change to $2,931.46, and the balance at the end of the year will change to $3,068.55. Repeating the above calculation will again show that the Total PV is the correct present value at day 0 to deliver the balance shown at the end of the year.

To look at a calendar year with 12 payments (no payment on the anniversary of the first payment), set Nr Pmts = 13 and then set the thirteenth payment to 0. You should see a total present value of $2,932.47 and a final balance of $3,082.50.

7.3   THE EFFECT OF INTEREST RATES ON PRESENT VALUE CALCULATIONS

The present value of amounts of money in the future, as the above calculations show, depends on the available interest rate. If interest rates are very low, then the present value looks essentially like the simple sum of all the payments. As interest rates get higher, payments far in the future become pretty worthless.

Table 7.3 shows the total present value of a 24 monthly payment account with $100 being deposited each month for different interest rates. As you can see from the table, the present value falls as the rate of interest increases.

One factor that's very easy to model on a spreadsheet but very difficult to predict in advance is just what interest rates will be a few years from now. As long as you can borrow money at a lower rate than you're paying on a loan, you stand to profit from the difference in rates.

To illustrate this, let's return to the simple example of the five-payment car loan shown in Table 7.1. You bought a car that's worth $4,546 in trade for five annual payments of $1,000. This is a perfectly equitable deal so long as prevailing interest rates remain at 5%.

Assume that high inflation occurs 3 years into your 5-year deal and savings bank interest rates jump to 10%. While high inflation in general is a terrible thing for the economy, in the case of this particular scenario, you can consider it a windfall. The money remaining in your account to pay off your car is now earning interest at a faster rate than anticipated while the amount pulled out each year to make a car payment, $1,000, doesn't change. This means that at the end of 5 years not only have you satisfied the payment agreement on your car, but that you've got money left in your bank account. On the other hand, if savings bank interest rates fall, your money is accruing interest more slowly than anticipated and there won't be enough in the account to make the last payment(s). You'll have to add some money out of your pocket to fulfill your obligation.

One last discussion point—the television sale example at the beginning of this chapter neglected some very important real-world considerations. The store owner borrowed money to buy the television set that he or she wants to sell to you. As long

Table 7.3   Present Value of the Example for Various Interest Rates

Interest rate (%)	Present value ($)
0	2,400.00
2	2,354.63
4	2,310.50
6	2.267.57
8	2,225.79
10	2,185.14
20	1,997.55

as that television is in his or her store, he or she cannot repay the loan and he or she is paying interest on the money. He or she really doesn't want to wait 2 years to sell you the set. In addition, new television models are coming out and in order to stay competitive, the store owner must be continually "turning over" his or her inventory, replacing older models with new models. On top of everything else, he or she has the fixed costs of renting his or her store from a building owner, paying his or her employees (and himself) a salary, and so on. Coming up with a pricing strategy that lets the store owner make money in this very dynamic situation is not a trivial task. Studying this in detail is far too involved a task for us here, but I'd like to emphasize that if you did set out to study it, the only concepts involved are those of compound interest and present value.

I'll present real-world situations where present value calculations are needed in the chapter on comparing loans and in the chapter on fixed annuities.

7.4 WHY THIS ALL MATTERS

You've probably picked up the answer to this already. The only way to compare costs (or values) of two or more transactions is to calculate these costs on the same day. In Chapter 8, I'll go through a detailed comparison of two mortgages to illustrate this point.

The bottom line is that if a transaction is not based on a correct value of the cost (the buyer) and the price (the seller), then one side is taking advantage of the other side. A correct value can be calculated on any date whatsoever, but all moneys coming and going must be translated to their value at that date. This was shown in the example above for the $5,000 loan.

7.5 A VERY INVOLVED EXAMPLE: WRITING YOUR OWN SPREADSHEET

As a last example, I'll present a calculation that probably can't be done easily using online calculators. In working through it on a spreadsheet, I'll also be able to easily extend our mathematical notation and take advantage of some spreadsheet mathematical capabilities that have not been introduced yet. If you are not comfortable with creating your own spreadsheets and don't want to venture down this path, you can skip this section.

Assume that you have access to an equity line of credit. That is, you have a checkbook that lets you write yourself loans that are backed by your equity in your home.

You want to buy a car on January 1 of the upcoming year, and you want to fund it 100% with a home equity loan. Your bank interest rate on this loan will be 8%, compounded monthly. Looking at your budget, you are pretty sure that you can make $250 a month payments. You want to fully pay off this loan in 4 years. Based on your work history, you're very confident that each year you will get a raise in pay, so that in the second year your can pay $300 a month; in the third year, $350 a

Table 7.4   Equity Loan with Unequal Payments

	A ($)	B	C ($)
1	000	1.000	0.0
2	250	1.007	248.34
3	250	1.013	246.70
4	250	1.020	245.07
5	250	1.027	1,217.21
6	1,250	1.034	241.83
7	250	1.041	240.23
8	250	1.048	238.64
9	250	1.055	237.06
10	250	1.062	235.49
11	250	1.069	233.93
12	750	1.076	697.14
13	000	1.083	00.00
14	300	1.090	275.17
46	400	1.349	296.62
47	400	1.358	294.66
48	900	1.367	658.59

month; and in the fourth year, $400 a month. You'd like to plan on several other variations; you usually spend a lot on presents during the holiday season in December, so you want to be able to skip January payments. Each year you get a tax refund sometime in May, so on June 1 you can add $1,000 to your regular payment. Right around Thanksgiving, your boss hands out bonuses, and you're confident that each December 1 you will be able to add $500 to your regular payment.

What color car should you buy?

Just kidding. The real question is how much of a loan can you take?

Table 7.4 shows a spreadsheet for working out this problem. The cell column numbers, on the left, are identically the month numbers for the loan. Row A shows the payment schedule. Row A must be entered by hand because the payments are the odd collection of numbers that I generated. If you add up all the payments, you get $20,300.00. At this point, you should already know that the present value of all these payments on day 1, which is the amount you can finance, will not be $20,300.00.

In column B, I want to put the amount that each of the payments in column A must be divided by to get the correct present value.

Call the monthly interest rate r. The present value of $250 paid 1 month from now is then

$$PV_1 = \frac{\$250}{(1+r)}.$$

The present value of $250 paid 2 months from now is just the present value of $250 paid 1 month from now, divided again by $(1 + r)$,

$$PV_2 = \frac{\$250}{(1+r)^2},$$

and then the present value of a $250 payment 10 months from now would just be written as

$$PV_{10} = \frac{\$250}{(1+r)^{10}}.$$

Spreadsheet notation is a bit different from textbook notation. Rather than typing $= (1 + r)^{10}$, in the spreadsheet you would type $= (1 + r)\char`^10$. It means exactly the same thing.

As an example, look at the tenth payment (row 10 in the spreadsheet). This is a $250 payment, so you would want to calculate $250/(1 + r)\char`^10$. This would work just fine, but if you do it this way, you have to type exactly what you want for each of the 48 payments. Instead, let's make full use of the power of the spreadsheet.

There is a spreadsheet function called "row" that returns the row of a selected cell. Let's say we want to put the number you divide the payment by to get the correct present value into column B. This calculation for the tenth payment would belong next to cell A10, the amount of the tenth payment, in cell B10. Working in cell B10, the function row(A10) returns the row that cell A10 is in (10). However, the tenth payment occurs 9 months after the start of the loan, so we need to enter row(A10) − 1.

Now let's put this all together. Remember that for an 8% annual loan, the monthly interest rate is 0.08/12, so in cell B10 you enter $= (1 + 0.08/12)\char`^(\text{row}(A10) - 1)$. If everything is correct, cell B10 should now say 1.062. Copy this formula up and down the entire B column (all 48 entries).

In row C, you'll put the present value of all the payments. Therefore, in C1, enter {=A1/B1} and copy this down row C for all 48 entries.

The number we're looking for, the sum of all 48 payments, is simply the sum of all of the entries in row C. Pick an empty cell and enter = sum(c1:c48). The term "sum" is another spreadsheet function. It simply adds up all the entries. The notation c1:c48 means "add up all the entries between c1 and c48." Your result should be $17,181.72.

A total of $17,181.72 is the amount you can borrow from your equity loan account and, if you meet the payment schedule you entered on the spreadsheet, after all 48 payments are made, the loan will be exactly paid off.

7.6 FUTURE VALUE

Analogous to the present value is the future value. Present value is the value today of some amounts of money that are known on different dates. Future value is the value at some future date, which must be specified, of some amounts of money that are known on different dates. The point here is that in various situations, you will be interested in the value of some amounts of money on some date and you must calculate, using known or at least estimated interest rates, what this value is. The only real difference is that present value means the value today (a unique date), whereas future value means the value at some date in the future that must be specified; "in the future" is not a unique date. Future value calculations are identically compound interest calculations, so I don't have to generate a new spreadsheet calculator. If I put $1 in the bank today at some APR and come back 10 years from today, my bank balance will be the future value of this $1 deposit.

7.7 PRESENT VALUE OF PREPAYMENT PENALTIES

In Chapter 5, I have presented several different possible prepayment penalties for loans, including the infamous rule of 78. I'd like to reexamine the last example of the chapter, that is, various prepayment penalties for a $300,000, 15-year loan, amortized monthly at 8% annual interest. Figure 5.3 shows the prepayment penalties versus the payment number when you decide to prepay the balance of the loan for four cases:

1. the rule of 78;
2. PP1: 6 months' interest on 80% of the balance;
3. PP2: 2% of the balance;
4. PP3: 3% of the balance for the first year, 2% for the second year, 1% for the third year, then 0 for the rest of the term of the loan.

In Figure 7.1, I redid Figure 5.3. I replaced each value with its present value at payment day 0 (the start of the loan) using the same 8% interest number as in the

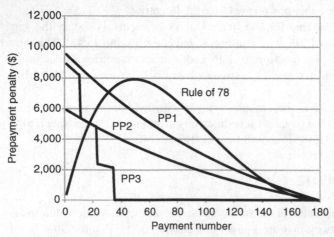

Figure 7.1 Present value of the prepayment penalties example in Chapter 5.

loan itself. I used the same horizontal and vertical axes for Figure 7.1 as I did for Figure 5.3 to make it easy to compare the two figures.

For low payment number values, the two figures look almost identical. This makes sense; the present value of an amount quoted only a few months away (at 8% annual interest) isn't much less than the quoted amount.

After about 60 payments, you start seeing the differences between the two figures. The worst prepayment penalty scheme, the rule of 78, peaks at about $8,000 rather than the almost $12,000 of Figure 5.3. Also, it peaks a bit sooner. Then, further out in time, all of the penalties (except PP3, which has already gone to 0), start falling very rapidly. At 10 years out (payment number 120), the amounts have dropped from about $3,000, $2,000, and $1,200 for rule of 78, PP1, and PP2, respectively, to about $6,000, $4,400, and $3,000. Compared with the balance of the loan at that time (about $139,500), these are getting to be not-too-significant nuisances.

PROBLEMS

Note: Since the principal purpose of Chapter 7 was to introduce the topic of present values so that later chapters could use this topic, there are very few problems below. Many problems will be presented in these later chapters that require calculating present values and using this information for various purposes.

1. I take a $22,000 auto loan, payable monthly at 6.00% APR for 5 years (60 payments). If I can save money at 4.00%, what is the present value of this loan the day that I take it?

2. A TV store dealer will "let me have" a $2,300 TV for 24 monthly payments of $100.00, the first payment due on the day I take the TV out of his or her store. Under what circumstances is this a good deal for me; under what circumstances is it a bad deal for me?

Chapter 8

Comparing Loans

In the 1981 movie *Absence of Malice*, Paul Newman played an honest businessman who is related to some organized crime figures. For various plot reasons, a newspaper publishes a story about him that contains true statements that are assembled to lead to a false impression/conclusion about his involvement in organized crime. The story ruins his reputation and his business. Most of the movie is involved with how he, using the same misleading techniques, gets even. Several times in the movie, somebody looks at a collection of facts pointing to a conclusion and asks, of the conclusion, "Is it true?" The answer is always, "No, but it's accurate."

Unfortunately, accurate but not true information is not just a clever ploy in an old movie plot. It is still with us today. Incorrect manipulation of loan details, although mathematically absolutely correct (accurate), nevertheless can lead to erroneous conclusions (but not true). The fake balance calculation that I have shown in Chapter 5 is a good example of this. Also, many times, the error comes from improper comparison of loans.

Before launching into calculations for comparing loans, I have to consider first why one loan is, in some sense, better than another loan. I'll limit myself to loans that are repaid with monthly payments such as a home mortgage loan or an automobile loan. There are many possible reasons for preferring one loan to another. If I'm short on cash, then the loan that offers the lowest possible payments looks best to me. I should consider how the lender will react if I'm able to increase my payments sometime in the future and what will happen if I'm able to pay off this loan early. The ability to pay off a loan early doesn't require that I win a lottery or that a long-lost great aunt dies and leaves me a fortune. Interest rates may drop significantly, and I might want to look at refinancing opportunities, possibly with a different lender.

If I know that I'm coming into a lot of money sometime relatively soon (I'm getting a signing bonus with a football team at the start of next season, or my trust fund pays me $1,000,000 when I reach age 25), then a low prepayment penalty is important.

Probably, the most common comparison criterion is the ultimate total cost of a loan. The total cost must be calculated either as the present value of the loan or the future value of the loan. That is, loans must be compared using their present values or their future values as of the same date. Present and future values are a little tricky

Understanding the Mathematics of Personal Finance: An Introduction to Financial Literacy, by Lawrence N. Dworsky
Copyright © 2009 John Wiley & Sons, Inc.

to assess because there must be some assumptions as to what interest rates will be at different times in the future. As Yogi Berra is reputed to have said, "It's very hard to make predictions, particularly about the future."

I'll propose several scenarios taking changing interest into account. In most cases, the better loan in a comparison will stay the better loan even if interest rates move around a bit.

Let's start with the most straightforward case: a mortgage loan with no up-front costs and a fixed interest rate for a fixed payment period. If two loans (for the same principal, of course) are amortized in the same number of payments, then clearly the loan with the lower interest rate must be the less expensive loan to take. On the other hand, what happens when two loans having the same principal have different interest rates and different payment periods, not to mention different savings bank (or CD) rates to use for the present value calculation?

The spreadsheet Ch8LoanPV.xls on my website will help to study this question. The first tab, labeled Basic, is very similar to the Mortgages spreadsheet on the website. The differences are that I eliminated the Month and Year entries, because all I want to track here are the number of payments, and that I added present value calculations.

The rate I used for the present value calculations is called the PV Rate. The variable Rate is the loan rate. In the example that's preloaded on this spreadsheet, I used Nr Pmts = 180; Principal = $350,000; Rate = 6.00%; and PV Rate = 3.00%. Tot PV is the sum of the present value of all the payments, that is, the present value of the loan, which in this example is equal to $427,682.80.

As I look at the present value of various loans, I'll keep the principal fixed at $350,000 for this example. The exact number for the principal doesn't matter, because everything will scale with this number. I want to vary the Rate, the PV Rate, and the NR Pmts. I'll create a table (Table 8.1), and then I'll discuss making a loan decision based on the data in this table.

The first thing that pops out at me when I look at Table 8.1 is that I've created a monster. It's just impossible to draw any general conclusions about these loans from this table. There are simply too many variables. Let's see if some common sense can narrow things down a bit.

When you look for a home loan, you usually get offers involving different interest rates, different payoff periods, and different up-front costs. I'll talk about factoring in up-front costs shortly. For now, let's consider the loan interest rate and the payoff period as "negotiable variables." You want the loan with the lowest PV that has monthly payments that you can handle.

Assume that interest on savings is about 3.50%. This lets me shrink the table significantly.

Table 8.2 is the rows of Table 8.1 that have a savings rate of 3.50%. I typed these items into a blank spreadsheet and then took advantage of the spreadsheet's ability to sort a table so that I've listed the items in increasing order of PV.[1]

---

[1] Virtually, all modern spreadsheet programs can sort. Where you find the command and just how you use it, however, varies from spreadsheet to spreadsheet so I can't give generic instructions. The Help function (usually the F1 key if it isn't on the menu) should get you going.

Table 8.1   Examples of Present Value of a Simple Mortgage Loan

Nr Pmts	Loan rate (%)	Savings rate (%)	PV ($)	Monthly payment ($)
120	6.00	3.00	402,412	3.886
240	6.00	3.00	452,131	2,508
360	6.00	3.00	497,725	2,098
120	6.00	3.50	392,950	3.886
240	6.00	3.50	432,359	2,508
360	6.00	3.50	467,309	2,098
120	6.00	4.00	383,793	3.886
240	6.00	4.00	413,794	2,508
360	6.00	4.00	439,539	2,098
120	7.00	3.00	420,854	4,064
240	7.00	3.00	489,285	2,714
360	7.00	3.00	552,309	2,329
120	7.00	3.50	410,958	4,064
240	7.00	3.50	467,885	2,714
360	7.00	3.50	518,558	2,329
120	7.00	4.00	401,382	4,064
240	7.00	4.00	447,794	2,714
360	7.00	4.00	487,743	2,329
120	8.00	3.00	439,771	4,246
240	8.00	3.00	527,867	2,928
360	8.00	3.00	609,144	2,568
120	8.00	3.50	429,431	4,246
240	8.00	3.50	504,783	2,928
360	8.00	3.50	571,920	2,568
120	8.00	4.00	419,424	4,246
240	8.00	4.00	483,108	2,928
360	8.00	4.00	537,933	2,568

Table 8.2   Subset of Table 8.1 for Savings Rate = 3.50% Sorted by PV

Nr Pmts	Loan rate (%)	Savings rate (%)	PV ($)	Monthly payment ($)
120	6.00	3.50	392,950	3,886
120	7.00	3.50	410,958	4,064
120	8.00	3.50	429,431	4,246
240	6.00	3.50	432,359	2,508
360	6.00	3.50	467,309	2,098
240	7.00	3.50	467,885	2,714
240	8.00	3.50	504,783	2,928
360	7.00	3.50	518,558	2,329
360	8.00	3.50	571,920	2,568

Looking at Table 8.2, the lowest PV and the highest PV are easy to guess; the shortest term loan at the lowest interest rate results in the lowest PV, while the longest term loan at the highest interest rate results in the highest PV. Generally, shorter term loans are better than longer term loans, and lower loan interest rates are better than higher loan interest rates. However, the actual order of all of the rows in the table isn't as obvious. It's the combination of both factors (loan interest rate and loan term) that produces the PV, not any single factor. If I had produced a table with smaller increments in both of these factors, there would be many more not-so-obvious results.

As the person about to take this mortgage loan, you now have a decision to make. The lowest PV loan (the top line in Table 8.2) is the best deal in terms of your actual total cost for the loan. What if you're uncomfortable, however, about making the (approximately) $3,900 monthly payment?

Table 8.2 gives you the information you need. The fifth line from the top shows you the loan with the lowest monthly payments. You didn't need this analysis to tell you that—the payment amount was certainly specified when the loan was offered to you. Nevertheless, let's assume that these nine loans were actually offered to you.

Rounding off to the nearest hundred dollars, the $2,100 monthly payment loan is the lowest payment loan. If you can afford $2,500 a month, go up the list to the fourth loan and take this loan. For no reason should you take the $2,300 a month loan (seventh on the list)—it will ultimately cost you more than the $2,100 a month loan.

Putting it all together, your best choice is the $3,900 a month loan, if you can afford it. Next best is the $2,500 a month loan and, finally, the $2,100 a month loan. If you can't afford to pay $2,100 a month, you'll have to go shopping for some other loans or, unfortunately, give up the idea of taking this mortgage loan—at least until interest rates drop.

8.1 UP-FRONT COSTS

The above discussion assumed simple loan conditions, including no up-front costs. Unless you've already factored the up-front costs into an effective loan interest rate, not taking these costs into account can lead to erroneous results.

Fortunately, the additional mathematics of taking up-front costs into account when calculating the PV of a loan is very simple because the PV of an amount due today (the initialization date of the loan) is simply the amount itself.

Returning to the spreadsheet, on the left-hand side is the entry Up Front. Any number put for this variable is added to Tot PV.

In terms of affecting the loan choice decision, up-front costs probably won't matter. Looking at Table 8.2, adding anything from $0 to $10,000 (different amounts to different loans) to the PVs won't change the order of things (unless the PVs are so close to start with that the difference is inconsequential).

If you want to fold the up-front costs into the loan, then your payments go up. In terms of comparing loans, however, it won't change the order of things unless the payment amounts were extremely close to start with. Consider the lowest payment loan, with a monthly payment of $2,098. Assume that the up-front costs are $13,000 (e.g., 3 points for $10,500 and closing costs of $2,500). As a fractional change in the amount of the loan,

$$\frac{\$13,000}{\$350,000} \approx 3.7\%.$$

Your monthly payment would therefore go up by about the same fraction, going to

$$\$2,098 + 0.037(\$2,098) = \$2,098 + \$78 = \$2,176.$$

8.2   ADJUSTABLE RATE MORTGAGES (ARMs)

It's impossible to accurately calculate the PV of an ARM because in the case of an ARM, not only don't we know the savings rate for the term of the loan, but we also don't know what the loan interest rate might change to (or how often). This doesn't mean however that we're completely lost when it comes to comparing loans. I can present a few plausible scenarios and see what we can learn.

Just to keep the number of variables more or less under control, I'll assume that the savings rate (PV annual percentage rate [APR]) on the spreadsheet is always one-half of the loan APR. Historically, this is a reasonable if not always an accurate assumption. Also, I'm leaving the up-front costs out of these calculations because, as I have just shown above, in almost all cases, they really don't change the final PV numbers, and hence the comparison process, in a meaningful way.

The ARM tab in spreadsheet Ch8LoanPV.xls (the same spreadsheet we've been using) is set up to calculate the PV of an ARM. Start on the left with the number of payments and the principal, then under Rate to the right of the green line (cell F2), enter the loan rate. The spreadsheet copies the loan rate down through all the payments and also sets the savings rate (PV Rate) to one-half of this number.

The interest paid per month numbers are shown in column 1.

I'll start with a $350,000 loan amortized over 15 years (180 payments) with a 5% interest rate. The PV interest rate will therefore start out at 2.5%. From either the Basic or the ARM tabs, you'll see that the monthly payment is $2,767.78 and the present value of the loan is $415,090 (rounded to the nearest dollar).

Now, suppose that after 5 years (60 payments), the adjustable rate kicks in and the loan rate goes up to 6%. Scroll down to payment number 60, and put the new rate into the corresponding Rate cell (F62). The new rate copies correctly, with cells changing the rate from the cell where you entered the new rate down to the bottom of the spreadsheet.

Looking at the payment column, you'll see that your payment has jumped to $2,897.08, starting with the sixty-first payment. This is a jump of about $210 a month, which your budget may or may not be able to comfortably tolerate.

Assuming that available savings rates go up proportionally with mortgage loan rates (the spreadsheet's default assumption), the new PV of the loan (at the start of the loan) is now $420,761. This is only about a $5,600 jump—barely more than 1% of the PV of the loan. If the available savings rate doesn't change (reset the PV Rate to 2.5% for the entire loan), then the PV jumps to $427,196. This is approximately a $12,100 change from the original PV, still just a small percentage change but starting to get noticeable.

What if your ARM doesn't change for the first 10 years of the loan? Reset the spreadsheet to the initial situation and then put the change to 6% after the 120th payment. After 10 years under the original terms of the loan, more than half of the loan is paid off (look at the balance column). The monthly payment only jumps about $70, and the PV jumps to $416,529. This latter jump is very small indeed.

Before I draw any conclusions from the above, I'd like to look at a few different scenarios. The numbers in Table 8.3 were all generated using this same spreadsheet. All of the loans are for $350,000 at an initial rate of 6% with an initial savings rate of 3%. I varied the term of the loan, the date when the mortgage gets adjusted, and the new mortgage loan rate. I kept the savings interest rate at half the mortgage loan rate—not because changing this wouldn't have generated interesting information, but because when too many things are changing, it gets very difficult to draw any conclusions and to develop any guidelines.

In Table 8.3, I've rounded all numbers to the nearest dollars and broken the data into four groups of three lines.

Table 8.3 ARM Loan Calculation Examples

Nr Pmts	Change Pmt Nr	Pmt at start ($)	PV at start ($)	New rate (%)	New Pmt ($)	% Change	New PV ($)	% Change
360	Fixed	2,098	497,725	6.00	2,098	0.00	497,725	0.00
360	60	2,098	497,725	7.00	2,302	9.70	512,618	2.99
360	120	2,098	497,725	7.00	2,271	8.22	507,495	1.96
240	Fixed	2,508	452,131	6.00	2,508	0.00	452,131	0.00
240	60	2,508	452,131	7.00	2,671	6.51	461,176	2.00
240	120	2,508	452,131	7.00	2,623	4.58	456,219	0.90
180	Fixed	2,954	427,683	6.00	2,954	0.00	427,683	0.00
180	60	2,954	427,683	7.00	3,089	4.58	433,275	1.31
180	120	2,954	427,683	7.00	3,025	2.42	429,104	0.33
360	Fixed	2,098	497,725	6.00	2,093	0.00	497,725	0.00
360	60	2,098	497,725	8.00	2,514	19.79	526,756	5.83
360	120	2,098	497,725	8.00	2,450	16.75	516,963	3.87

Looking first at the fixed interest entries reasserts what we already know. That is, taking the same loan for a longer term results in lower payments but a higher ultimate cost (PV) of the loan.

Look at the first three groups of entries. These are all ARM loans that jump from 6% to 7% after 60 months (5 years) or 120 months (10 years). In every case, the change in interest rate results in a larger payment, but the percent change in payment is smaller for the shorter term loans. The same conclusion applies to the PV of the loans. That is, the PV goes up, but the percent change in PV is less for shorter term loans.

Before taking a loan, you must compare starting interest rates as well as loan terms, but these numbers seem to be telling us that you want to take as short a term loan as you can (without sacrificing initial interest rate too much) and come up with as large a payment as you can within your budget.

In terms of what happens when the mortgage gets adjusted up a percentage point or two, these numbers are telling us that the further out you can push the change, the better off you are. A 10-year fixed percentage ARM is better than a 5-year fixed percentage ARM. Also, note that while the percent change in payment can be upsettingly high, the percent change in the PV—the actual cost of the loan—doesn't seem to jump that much, especially for the 10-year ARMs.

You should use this spreadsheet to compare actual loans that you can get when you're ready to buy your house. Read the loan terms and understand just how much and when the interest rate might jump, and then create your own table from the spreadsheet using severe case if not worst case scenarios. Remember that you can change the interest rate as many times as there are payments on the spreadsheet, but be careful to undo or reinitialize everything carefully when you want to move on to another loan.

8.3   A FEW LAST WORDS

When interest rates go down, it's time to investigate opportunities for refinancing your home. This means that you look for a new mortgage loan that will replace your old mortgage loan at a lower interest rate. In many cases, the up-front costs of the new loan are very low (sometimes even 0). As always, folding the up-front costs into the new loan gives you an effective interest rate and a resulting monthly payment number.

There are some "gotchas" in refinancing a home that you have to keep in mind. Let's say that you are 15 years into a 30-year mortgage and the opportunity comes along to refinance at a lower rate. If the new loan is a 15-year loan, then you'll see that your monthly payments drop and that your home is paid off 15 years after financing, which is when the original (30-year) mortgage would have been paid off. If you take a new 20- or 30-year mortgage loan, your payments will drop even further, but you might be signing a contract that requires you to make payments well into your intended retirement. Work through the numbers and make sure you want to do this; don't be seduced by the lower monthly payments.

Using the same scenario as above, after 15 years of payments, you've built some equity[2] in your home, and in good economic times, home values usually increase over time, further building equity in your home. This gives you the opportunity to refinance for more money than you owe on your first mortgage and therefore to "walk away" with some money.

Whether or not doing this is a good idea depends on just how you use the money. My point here is that you should always work through the numbers and really understand all sides of what you're doing when you make major financial decisions.

A few paragraphs ago I slipped something by that is true but can be looked at in a number of ways depending on your perspective. Consider a fixed rate, 30-year mortgage loan taken for $350,000 at 8% interest. The monthly payment is $2,568.18. At the end of 15 years, the outstanding balance is $267,959. Let's say you refinance and take a new 15-year loan for $300,000, giving yourself almost $32,000 in cash to use as you wish. This doesn't mean that your monthly payments will now be higher than they used to be. If you can get the new loan for 6.23%, your new payments are (to within about a dollar) the same as the old payments. If you can get the new loan for less than 6.23%, then your payments are lower than they used to be.

You might think of this $32,000 as "free money" because you continue making the same monthly payments that you've always made, with the loan being fully paid off at the same time as originally planned. However, when you refinanced, the equity in your home dropped by the $32,000 that you took away as cash. This $32,000 drop will diminish gradually for the next 15 years, reaching 0 at the last payment of the loan. However, if you sell your house sometime during this second 15-year period, you will walk away with less money than if you had never taken the $32,000.

In the next chapter (Chapter 9), I will discuss about the effects of taxation and inflation on savings and long-term loans such as mortgages. Very briefly here, the principal effect of taxation on a mortgage loan for your home is that, at least today and with some qualifying details, the interest is deductible from your federal (IRS) taxes. If you can estimate your taxable income, you can estimate a lower effective interest rate for your home mortgage.

Inflation, on the other hand, raises the prices of things that you need to or want to buy, thereby making your dollar less valuable. This is bad news for how much your savings will be able to buy in the future. If you're lucky enough to have a job or a business that allows your income to climb along with inflation, you are paying off your home mortgage with cheaper dollars (i.e., a smaller percentage of your salary) every year.

PROBLEMS

1. Consider a fixed rate, fixed payment 15-year mortgage loan for $350,000. I'll consider the present value of the loan when I take the loan to be my actual cost of the loan. If I want

---

[2] Your equity in your home is the resale value of the home minus all outstanding debts. In other words, if you were to sell your home and pay off your mortgage loan(s), your equity is the amount of cash left in your hand.

to keep this present value constant at about $400,000, how much can I borrow based upon the APR of the loan? Assume the savings interest rate is one-half of the loan APR.

2. Find the monthly payments for the above situation, using APRs of 0%, 2%, 4%, 6%, 8%, and 10%. Then divide each monthly principal by its accompanying payment, giving you the ratio of principal: payment.

3. In this chapter, I argued that the best choice for a loan is the loan with the lowest present value, subject to the constraint that you have to be able to afford the payments. We could consider a figure of merit (FM) for a loan as the product of the present value and the monthly payment—when both of these factors are low, the FM will be low; when both are high, the FM will be high; and intermediate cases will fall somewhere in between. Using the data in Table 8.2, calculate the FM for each loan and discuss whether or not this FM definition is good.

   In the table, I got the FMs by multiplying the PV by the Monthly Payment and then dividing by 10,000,000. These divisions don't change the relative values of the FMs, but they do make it a lot easier to read the significant figures of the numbers without getting distracted by all of the digits.

Nr Pmts	Loan rate (%)	Savings rate (%)	PV ($)	Monthly payment ($)	Figure of merit
120	6.00	3.50	392,950	3,886	
120	7.00	3.50	410,958	4,064	
120	8.00	3.50	429,431	4,246	
240	6.00	3.50	432,359	2,508	
360	6.00	3.50	467,309	2,098	
240	7.00	3.50	467,885	2,714	
240	8.00	3.50	504,783	2,928	
360	7.00	3.50	518,558	2,329	
360	8.00	3.50	571,920	2,568	

4. You have a 15-year, $350,000 mortgage with a 5.00% APR. Eight years into the mortgage, you get the opportunity to refinance with a new 8-year loan for the remaining balance of your loan at 4.2%. The up-front cost for this refinancing is $3,000. Is this a good deal or not?

5. I want to buy a new home and I need to borrow $500,000. Lender A has a very attractive APR of 5.10% on a fixed 30-year mortgage, but he or she will only lend me $350,000. Lender B will give me a second mortgage for the remaining $150,000, but he or she wants 7.20% for a fixed 30-year mortgage. Lender C will loan me the entire $500,000, but he or she wants 6.5% for a fixed 30-year mortgage. What should I do?

   Your first instinct should be to treat this as a trick question. The best financial deal is to take Lender A's loan for $350,000 and then take only the remaining $150,000 from lender C. I'm going to force you to do some work here: Lender C handles "jumbo" mortgages only; $500,000 is his or her minimum mortgage—take it or leave it.

Chapter 9

Taxation and Inflation

John Lennon once said, "Life is what happens to you while you're busy making other plans." Both taxation and inflation, it seems, are parts of life—they happen to you while you're busy making other plans.

Taxation is the government's way of getting money to pay its bills. You pay taxes on, among other things, your income; earned interest may be considered part of your income. The government (federal, state, and local) all want a piece of this income, so you get to keep and spend or save less than what all of the calculations thus far have promised you. On the other hand, interest on some of your debts is considered to be a deductible expense. That is, you get to reduce the income you report to the government, from which your tax burden is calculated, by this interest.

Inflation is not a phenomenon that takes money away from or adds money to your bank account. Instead, inflation reduces the spending power of your money. Out of control inflation rates can be disastrous to a country's economy, so the government watches it and tries to control it very carefully. There are circumstances where inflation can work for you. These circumstances will be discussed later on in this chapter.

9.1 UNDERSTANDING PERSONAL FEDERAL INCOME TAX RATES

Every year in the United States, on April 15, our income tax filings are due. The federal government, most states, and some cities collect income taxes. In almost all cases, the federal tax is by far the highest of the personal income taxes that we must pay.

When you fill out your tax return, you list all of your earnings and then you list your deductions. When you subtract the sum of all your deductions from the sum of your earnings, you get your *taxable income*. The amount of tax you must pay to the Internal Revenue Service (IRS) is based on this taxable income and your *filing status*. Your filing status can be single, married filing jointly, and so on. Accompanying each filing status is a Tax Rate Schedule.

Understanding the Mathematics of Personal Finance: An Introduction to Financial Literacy, by Lawrence N. Dworsky
Copyright © 2009 John Wiley & Sons, Inc.

2008 Tax Rate Schedules

Schedule X—If your filing status is Single

If your taxable income is:		The tax is:	
Over—	But not over—		of the amount over—
$0	$8,025	.... 10%	$0
8,025	32,550	$802.50 + 15%	8,025
32,550	78,850	4,481.25 + 25%	32,550
78,850	164,550	16,056.25 + 28%	78,850
164,550	357,700	40,052.25 + 33%	164,550
357,700	....	103,791.75 + 35%	357,700

Schedule Y-1—If your filing status is Married filing jointly or Qualifying widow(er)

If your taxable income is:		The tax is:	
Over—	But not over—		of the amount over—
$0	$16,050	.... 10%	$0
16,050	65,100	$1,605.00 + 15%	16,050
65,100	131,450	8,962.50 + 25%	65,100
131,450	200,300	25,550.00 + 28%	131,450
200,300	357,700	44,828.00 + 33%	200,300
357,700	....	96,770.00 + 35%	357,700

Schedule Y-2—If your filing status is Married filing separately

If your taxable income is:		The tax is:	
Over—	But not over—		of the amount over—
$0	$8,025	.... 10%	$0
8,025	32,550	$802.50 + 15%	8,025
32,550	65,725	4,481.25 + 25%	32,550
65,725	100,150	12,775.00 + 28%	65,725
100,150	178,850	22,414.00 + 33%	100,150
178,850	....	48,385.00 + 35%	178,850

Schedule Z—If your filing status is Head of household

If your taxable income is:		The tax is:	
Over—	But not over—		of the amount over—
$0	$11,450	.... 10%	$0
11,450	43,650	$1,145.00 + 15%	11,450
43,650	112,650	5,975.00 + 25%	43,650
112,650	182,400	23,225.00 + 28%	112,650
182,400	357,700	42,755.00 + 33%	182,400
357,700	....	100,604.00 + 35%	357,700

Figure 9.1   IRS 2008 Tax Tables.

Figure 9.1 shows the 2008 Tax Rate Schedules. Each schedule is a table. Each row in the table shows a range of taxable income and a formula for calculating the tax for this income. Each formula includes a percentage that is known as the *tax bracket*. What your tax bracket does and doesn't tell you about your tax and its relation to your taxable income are often misunderstood. I'll point out the place where many people get it wrong as I walk through the calculations.

Table 9.1 is a repeat of Figure 9.1 for married couples filing jointly with the tax table formulas rewritten in conventional algebraic notation. As an example, if your taxable income is exactly $100,000, then your tax is

$$\text{Tax} = \$8,962.50 + 0.25(\$100,000 - \$65,100) = 8,962.50 + 0.25(\$34,900)$$
$$= \$8,962.50 + \$8,725.00 = \$17,687.50$$

In Figure 9.2, I've plotted the tax versus the taxable income for married couples filing jointly. If you look carefully, you can see that the graph is made up of straight

Table 9.1 IRS 2008 Tax Tables for a Married Couple Filing Jointly

Taxable income range (x)	Tax
$\$0 < x \le \$16,050$	$0.1x$
$\$16,050 < x \le \$65,100$	$\$1,605.00 + 0.15(x - \$16,050)$
$\$65,100 < x \le \$131,450$	$\$8,962.50 + 0.25(x - \$65,100)$
$\$131,450 < x \le \$200,300$	$\$25,550 + 0.28(x - \$131,450)$
$\$200,300 < x \le \$357,700$	$\$44,828 + 0.33(x - \$200,300)$
$\$357,700 < x$	$\$96,770 + 0.35(x - \$357,700)$

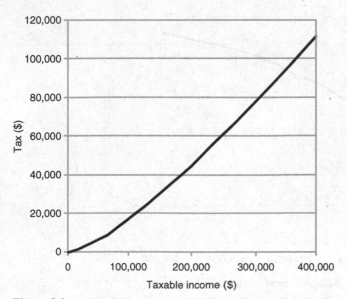

Figure 9.2 Graph of IRS 2008 Tax Tables for a married couple filing jointly.

line segments with slight corners where they join. These corners occur when the taxable income jumps from one tax bracket to the next.

The fact that the graph is continuous—there are no discontinuous jumps in the graph—is important. While it is clear that the higher your taxable income is the higher your taxes will be, there are no sudden jumps when you move from one tax bracket to another. In other words, if you earn a few dollars more, you pay a few dollars more. Somehow, the idea has snuck into the popular culture, that if you earn a few dollars more and happen to move to a higher tax bracket you will suddenly owe a huge amount more in taxes. There is simply no basis for this conclusion.

Because Figure 9.2 is not a straight line, the average tax rate is not a single number. The average tax rate on your taxable income is your tax divided by your taxable income, usually expressed as a percentage. If you remember the graphs discussion in Chapter 1, this is identically the slope of the line from the lower left-hand corner of the graph to the point on the graph showing your taxable income and tax.

Figure 9.3 shows this calculation, with the average tax rate plotted versus the taxable income (for married couples filing jointly). The graph has some irregular curves in it, but it is monotonically increasing. This means that as your taxable income (the horizontal axis) increases, your average tax rate increases (the vertical axis). The details of how it increases depending on your taxable income bear some discussion, but nonetheless, it always increases.

Using Figure 9.3, let's consider a taxable income of $100,000 as an example. The average tax rate on these earnings is about 18%. Your tax bill will be 18% of $100,000, or $18,000. Looking back at the tax table, however, a taxable income of $100,000 puts you in the 25% bracket. If you're only paying 18% of your taxable earnings as tax, how does this have anything to do with 25%?

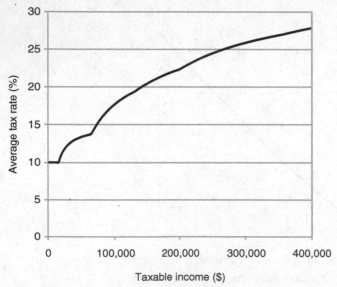

Figure 9.3 Average tax rate for a married couple filing jointly.

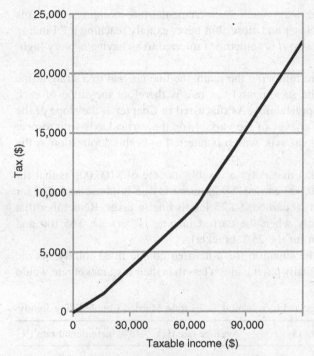

Figure 9.4   Expansion of the lower left region of Figure 9.2.

The answer is that the term "tax bracket" refers to the incremental tax rate, not the average tax rate. Figure 9.4 shows the expanded lower left corner of Figure 9.2. The graph is actually made up of connected straight line segments, each with a different slope. These line segments connect at the break points between the tax brackets. When your taxable income is $100,000, the first $16,050 is taxed at a 10% rate; $65,100 − $16,050 = $49,050 is taxed at a $15% rate; and then $100,000 − $65,100 = $34,900 is taxed at a 25% rate. The average tax rate comes from averaging all of these pieces together. This is called a *progressive* tax. The name has nothing to do with the political party or philosophy that created it. Progressive here refers to the fact that the incremental tax rate gets progressively larger as the taxable income increases.

Your average tax rate only gets close to your incremental tax rate only when you're making an awful lot of money. Using the same tables as above, you can see that once your taxable income is more than $357,700, you are in the top tax bracket. For a $400,000 taxable income (the largest taxable income that I included in the graphs), your average rate is about 28% while your incremental rate is 35%. Dividing the former by the latter, we get that your average rate is 80% of your incremental rate.

If your taxable income was $1,000,000, then your tax would be about $322,000. Your average tax rate then is 32%, but your incremental rate is still 35%. Your average rate is now 92% of the incremental rate. At $2,000,000 taxable income, your average rate would be 99.6% of your incremental rate, effectively the incremental rate. The average rate is said to approach the incremental rate *asymptotically*. This is a fancy term for "getting closer and closer but never exactly reaching it." Finding yourself in this situation, however, is sometimes referred to as having a "very high-class problem."

For those of you who are following the math, the line tangent to a straight line is the straight line itself. The incremental tax rate is therefore the slope of each straight line in its range of applicability. As discussed in Chapter 1, the slope of the line is also called the rate of change of the variable on the vertical axis with respect to the variable on the horizontal axis, which is referred to in this application as the "tax bracket."

What the 25% tax bracket means for a taxable income of $100,000 is that for every dollar more than $100,000 earned, $0.25 more is due in taxes, and also for every dollar less than $100,000 earned, $0.25 less is due in taxes. Remember that this last sentence applies only when the earned income is between $65,100 and $131,450 (the defined region for the 25% bracket).

Table 9.2 summarizes the situation for a married couple filing jointly. As the table shows, everybody is actually paying much less than their tax bracket rate would

Table 9.2 Summary of Average and Incremental Taxes for a Married Couple Filing Jointly

Taxable income ($)	Tax ($)	Average rate (%)	Incremental rate (%)
0	0	0	0
10,000	1,000	10	10
20,000	2,198	11	15
40,000	5,198	13	15
60,000	8,198	14	15
80,000	12,688	16	25
100,000	17,688	18	25
120,000	22,688	19	25
140,000	27,944	20	28
160,000	33,544	21	28
180,000	39,144	22	28
200,000	44,744	22	30
220,000	51,329	23	33
240,000	57,929	24	33
260,000	64,529	25	33
280,000	71,129	25	33
300,000	77,729	26	33
320,000	84,329	26	33
340,000	90,929	27	33
360,000	97,575	27	35
380,000	104,575	28	35
400,000	111,575	28	35
20,000,000	6,971,575	34.9	35

predict. Even with a taxable income of $20,000,000, you're not quite paying your tax bracket rate—but you are getting very close.

9.2 ONLINE TAX CALCULATORS

Many online calculators will actually calculate your taxable income from the information you give them. Make sure you know what's being calculated, based on what. These links each lead to several calculators and tax information:

1. http://turbotax.intuit.com/tax-tools/;
2. http://www.dinkytown.net/java/Tax1040.html;
3. http://www.hrblock.com/taxes/tax_calculators/index.html;
4. http://www.finance.cch.com/sohoApplets/index.html (This link brings you a large list of calculators. Scroll down the list to find the tax calculators.)

9.3 TAXATION OF EARNED INTEREST

While determining what interest is taxable and what interest isn't taxable can be a fairly sophisticated task, figuring out the tax on taxable interest is very straightforward. Let's say that your taxable income is $50,000 not counting interest. Your tax is approximately $6,700, which is an average tax rate of 13% and an incremental tax rate of 15%. If you have $10,000 of savings earning 4% interest, you earned $400 in interest last year. This interest is taxed at the 15% rate, so the tax on your interest is 0.15(400) = $60.

If your interest was $400 but you had to send $60 back to the government, then you only got to keep $400 − $60 = $340. Your effective interest rate was therefore

$$\$340/\$10,000 = 3.4\%.$$

If your taxable income, not counting interest, had been $360,000, your tax would have been approximately $97,600. This is an average tax rate of 27% and an incremental tax rate of 35%. Now, the tax on the $400 interest is 0.35(400) = $140, and you get to keep only $260 of the $400. Your effective interest rate on your earnings is

$$\$260/\$10,000 = 2.6\%.$$

Remember that the tax on your earnings excluding this interest does not go up because of the interest earnings. It remains exactly the same.

Some very secure investments are "tax free." This means that you do not pay taxes on the interest from these investments. This tax-free status might apply to the federal tax, the state tax, or both—research this point carefully. If one of these investments paid 3% interest when you could get 4% interest on a certificate of deposit, you might first head toward the certificate of deposit. If you are in a high enough tax bracket that the effective interest rate on the certificate of deposit is lower than 3%, as it is in the above example, then the tax-free investment suddenly starts to make sense.

Calculating the tax and the effective interest rate on an investment gets a little bit more complicated if your taxable income is very close to (but less than) the upper edge of a tax bracket.

For example, suppose your taxable income was $130,000 and you earned $3,000 in interest last year; staying with the married filing jointly table, note that the tax bracket jumps from 15% to 25% at a taxable income of $131,450. Therefore, $1,450 of your interest gets taxed at 25% and the remaining $1,550 gets taxed at 28%:

$$\text{Tax} = 0.25(\$1,450) + 0.28(\$1,550) = \$362.50 + \$434.00 = \$796.50.$$

9.4 DEDUCTIBLE INTEREST

The calculation of the amount of tax you don't have to pay if you can deduct some interest is the same calculation as for the tax on taxable interest; you subtract the results from your taxable income rather than adding the results to it.

For example, suppose your taxable income is $100,000, which (same tax table as above) would result in your owing approximately $17,700 in taxes. If you have a sizeable loan such as a home mortgage, you might have paid $10,000 in interest over the course of the year. If you can list this interest as a deduction, your taxable income drops to $90,000 and your tax drops to approximately $15,200. This is a saving of $17,700 − $15,200 = $2,200 in tax.

Assuming that your loan is at 6% interest, we can approximate an effective interest rate by imagining that we're repaying the tax savings to the loan, and therefore

$$Eff\ Int\ Rate = \frac{\$10,000 - \$2,200}{\$10,000}(6\%) = 4.68\%.$$

9.5 DEFERRED TAXATION SAVINGS

The U.S. government offers several different tax deferral schemes for workers to use to build their "nest egg." There are 401K plans, individual retirement accounts (IRAs), Roth IRAs, self-employed pension (SEP) IRAs, and so on. Each of these plans serves a different purpose/customer base, and each has its own rules about taxation, contribution, withdrawal, and so on. Many books have been written on this topic, and I couldn't possibly do it justice in a few pages. What I will do in the following pages is to walk through a hypothetical saving for a retirement example and show the potential value of a tax-deferred savings plan. I will be using my online spreadsheet Ch9Taxation.xls.

I'll begin using the Nest Egg tab on the spreadsheet. You are 40 years old, planning to retire at age 65. Your income puts you well into the 25% tax bracket.

Table 9.3   Nest Egg Building Example

Input variables: Nr Mnthly Pmts = 300
Interest rate = 4.00%
Mnthly Pmt = $250
Tax rate = 25%

Monthly contribution after withholding tax: $188

Pmt Nr	Month	Year	Taxed savings				Tax-deferred savings	
			Balance ($)	Interest ($)	Tot Int/ Year ($)	Tax ($)	Balance ($)	Interest ($)
1	1	1	188	0	0		250	0
2	2	1	376	1	1		501	1
3	3	1	564	1	2		753	2
4	4	1	754	2	4		1,005	3
5	5	1	944	3	6		1,258	3
6	6	1	1,134	3	9		1,513	4
7	7	1	1,326	4	13		1,768	5
8	8	1	1,518	4	18		2,023	6
9	9	1	1,710	5	23		2,280	7
10	10	1	1,903	6	28		2,538	8
11	11	1	2,097	6	35		2,796	8
12	12	1	2,281	7	42	10	3,056	9
13	1	2	2,476	8	8		3,316	10
14	2	2	2,672	8	16		3,577	11
15	3	2	2,869	9	25		3,839	12
296	8	25	82,737	274	2,151		125,841	417
297	9	25	83,201	276	2,427		126,511	419
298	10	25	83,666	277	2,704		127,182	422
299	11	25	84,132	279	2,983		127,856	424
300	12	25	83,784	280	3,264	816	128,532	426

You want to divert $250 of your gross salary each month to an account that will give you 4% interest annual percentage rate (APR). This example is shown in Table 9.3.

The first thing that happens is that some money is withheld from your paycheck for taxes. To keep things simple, I'll assume that the exact tax payment is deducted each month. I'm ignoring social security, Medicare, state tax, and any other deductions. The money you have available to contribute is 75% of $250 = $188 (cell K1 on the spreadsheet). The spreadsheet shows the balance growth each month and the interest paid into the account each month. At the end of the year, tax is paid on the year's interest, and this amount is deducted from the balance.[1]

At the end of 25 years (300 monthly payments), your balance, after taxes, is $83,784.

Look at the tax-deferred savings columns. Assuming your plan deposits payments pretax (some, but not all of them do), the full $250 goes into the account every month. The interest accrued does not get taxed, and at the end of the 25-year period, the balance in the account is $128,532.

The tax-deferred plan has about 50% more in it than the simple savings plan. Remember, however, that the simple savings plan money is all yours at this

Table 9.4   Retirement Phase of Nest Egg Building Example

Input variables: Monthly withdrawal = $750
Tax rate = 10%
Interest rate = 4.00%

	Taxed savings				Tax-deferred savings		
Month	Balance ($)	Interest ($)	Tax ($)	Month	Balance ($)	Interest ($)	Tax ($)
1	83,382	278	28	1	127,782	426	75
2	82,882	276	28	2	127,383	425	75
3	82,381	275	27	3	126,983	423	75
4	81,878	273	27	4	126,581	422	75
5	81,374	271	27	5	126,178	421	75
6	80,868	270	27	6	125,774	419	75
7	80,360	268	27	7	125,368	418	75
8	79,851	266	27	8	124,961	417	75
9	79,341	264	26	9	124,552	415	75
134	1,832	6	1	134	61,130	204	75
135	1,087	4	0	135	60,509	202	75
136	341	1	0	136	59,886	200	75
137	−408	−1	0	137	59,260	198	75
				138	58,633	195	75
				139	58,003	193	75
				217	1,842	6	75
				218	1,024	3	75
				219	202	1	75
				220	−622	−2	75

[1] If I were doing present value calculations, I'd note that the tax on the year's interest doesn't get paid until the following April. For these examples, I'm keeping things simple by just deducting the tax payment from the balance at the end of the year.

point—it's all after-tax money. The money in the deferred tax plan is still yet to be paid.

Now look at the Retirement tab of Ch9Taxation.xls (shown in Table 9.4). I've simplified the calculations in that I show the tax generated each month being paid that month. Let's say you would like to draw $750 each month from your account. Since you're retired, you're probably in a lower tax bracket than you were when you were working. I'll use 10% for the example. I left the interest rate at 4%; this of course could have changed over the years.

In the case of the taxed savings account, the tax is the tax rate multiplied by the interest earned (each month). For this tax-deferred plan, you pay taxes on the money as you withdraw it. The taxed savings monthly tax starts off lower than the tax-deferred savings monthly tax and gets a little lower every month. The higher initial balance in the tax-deferred savings account outweighs this tax disadvantage, however: The taxed savings account runs out of money in about 11½ years, while the tax-deferred account lasts about 18½ years. This is significantly a better retirement funding.

A word of warning: Each government-approved plan has its own rules about taxing contributions and taxing disbursements. There are also rules about withdrawing money from the account while still in the "nest egg" phase, as well as rules about maximum contributions, and rules about everything else (there are probably rules about writing rules). To make it worse, these rules keep changing. Do your homework.

In good times, many employers will match employees' contributions to retirement funds up to some maximum percentage specified by the IRS. This is an effective pay raise. If you find yourself in this fortunate situation, increase the nest egg contribution amount in the spreadsheet by your employer's contribution to see how your balance will grow. Keep in mind that this contribution only applies to the tax-deferred savings numbers; there is no matching contribution for a personal savings account.

9.6 ONLINE DEFERRED TAXATION PLAN CALCULATORS

This site calculates how a 401K savings plan benefits you: http://www.bloomberg. com/invest/calculators/401k.html.

The bloomberg.com host site is a treasure trove of financial calculators: http:// www.bloomberg.com/invest/calculators/index.html.

This site isn't a calculator. It's a U.S. government (Department of Labor) site about 401K plan fees: http://www.dol.gov/ebsa/publications/401k_employee.html.

9.7 INFLATION

When you pay taxes or get a tax deduction, you can see the actual dollar amounts coming and/or going, relate this to expenditures and income, and plan your financial activities accordingly. Inflation is different. Inflation is a devaluation of the buying power of a dollar. Assume that you have a savings account that earns 4% a year interest and you deposit $1,000 into it at the beginning of the year. At the end of the year, you would have $1,040 in your savings account and you should be able to buy more with this $1,040 than you could have bought with your original $1,000. If you left the money, at the end of the second year you would have $1,081.60 (it should be obvious that I'm compounding annually), and you should be able to withdraw this money and buy even more if you so wished. If you want to buy bricks that cost $1.00 each, when you had $1,000 you could have bought 1,000 bricks. After the first year, you could have bought 1,040 bricks, or after the second year, you could have bought 1,082 bricks (rounding the numbers a bit).

Suppose that inflation is running at 6% a year. Starting with $1,000, when bricks are $1 each, you can buy 1,000 bricks. After the first year, you have $1,040 in your bank account, but the same bricks now cost $1.06 each. You can only buy 981 bricks. If you leave your money in the bank for 2 years, you have $1,082, but bricks now cost $1.12 each and you can only buy 966 bricks. Even though your bank account is growing, you are effectively getting poorer every year.

Inflation is particularly hard on people with fixed incomes, for example, retired pensioners. Each month, they receive the same check, and each month it's worth less. The U.S. government provides periodic cost of living adjustments to social security payments to help counteract this problem. Go to http://www.ssa.gov/OACT/COLA/latestCOLA.html to learn the specifics of this legislation.

Out of control inflation can destroy a country. Germany, after the first world war, had this problem; historians strongly correlate the high inflation rates to the demise of the Weimar Republic and to the rise of the Nazi party to power.

A low inflation rate that does not change abruptly is not difficult to live with. Savings bank interest rates tend to be higher than the inflation rate, so that actual savings and growth of buying power are possible.

Factoring the reality of inflation into long-term saving and borrowing is important. In the example I'm about to present, I'm using numbers that were chosen to be convenient for creating an example. I don't know what the years will bring.

For this example, I'll assume that inflation is a never-changing 4%. On January 1, 2000, I took a job that, after taxes, left me $48,000 a year to live on. Using existing savings for a down payment, I bought a home with a mortgage loan of $2,000 a month, or $24,000 a year, payments for 30 years. This might have been too much of a loan because my monthly mortgage loan payments eat up half of my available funds, but I am frugal and live within my means.

The company I work for is able to raise the price of its products to track with inflation; its costs also track with inflation, and even though I don't advance much in my career, my salary also tracks with inflation. This means that in 2020, my available spending money, after taxes, is about $105,000.[2] My grocery, gasoline, and so on costs also scale with inflation so it would appear that my standard of living has not changed.

One thing that did not scale with inflation is my monthly mortgage payment. I have a fixed rate, fixed payment, mortgage so I am still paying $24,000 a year on

---

[2] I'm neglecting a phenomenon called "bracket creep." Even though your salary might just be tracking inflation, if the tax laws aren't updated, then you slowly move into higher tax brackets and your tax goes up. You are, in effect, getting poorer.

my mortgage loan. But look what has changed: My mortgage loan payments used to be 50% of my after-tax earnings. In 2020, however, they are $24,000/$105,000 ~ 23% of my after-tax earnings. My standard of living certainly has changed—I now have more than 75% of my after-tax earnings to spend. I'm also well along the way to paying off my home, and if the next 10 years is like the first 20 years, it will get easier to make the payments every month.

Real situations are much more complicated than my simple examples.

To illustrate the impact of even modest inflation on planning for retirement (or any fixed income situation), I added an Inflation tab to the Ch9Taxation.xls spreadsheet. Hopefully, over the nest egg period, your salary was growing and you were able to periodically raise your contributions to your accounts. I'll redo the retirement calculations with inflation to account for (Table 9.5).

Table 9.5   Retirement Phase of Nest Egg Building Example with Inflation

Input variables: Monthly withdrawal = $750
                   Tax rate = 10%
                   Interest rate = 4.00%

	Taxed savings				Tax-deferred savings			
Mo	Withdraw ($)	Balance ($)	Int ($)	Tax ($)	Mo	Withdraw ($)	Balance ($)	Int ($)	Tax ($)
1	750.00	83,382	278	28	1	750.00	127,782	426	75
2	751.25	82,881	276	28	2	751.25	127,382	425	75
3	752.50	82,377	275	27	3	752.50	126,979	423	75
4	753.76	81,870	273	27	4	753.76	126,574	422	75
5	755.01	81,361	271	27	5	755.01	126,166	421	75
6	756.27	80,849	269	27	6	756.27	125,755	419	75
7	757.53	80,334	268	27	7	757.53	125,342	418	75
8	758.79	79,816	266	27	8	758.79	124,926	416	75
9	760.06	79,295	264	26	9	760.06	124,507	415	75
119	912.85	2,182	7	1	119	912.85	59,507	198	75
120	914.38	1,275	4	0	120	914.38	58,716	196	75
121	915.90	363	1	0	121	915.90	57,921	193	75
122	917.43	−554	−2	0	122	917.43	57,122	190	75
					123	918.96	56,318	188	75
					124	920.49	55,511	185	75
					180	1,010.46	3,019	10	75
					181	1,012.14	1,942	6	75
					182	1,013.83	859	3	75
					183	1,015.52	−228	−1	75

Comparing Table 9.5 to Table 9.4, I have added a withdrawal column, which indexes the amount of money withdrawn up with inflation. I am assuming here that your cost of living will go up directly with inflation. This will not be the case if you have a fixed interest rate mortgage loan. In this case, only part of your expenses will go up with inflation. To correct for this, in the spreadsheet, just scale back the anticipated rate of inflation to account for the fraction of your cost of living that will not grow with inflation.

Comparing the two tables, the savings bank account runs out of money 15 months earlier and the tax-deferred account runs out of money 17 months earlier due to inflation-indexed withdrawals. The tax-deferred account, with or without inflation issues, still outperforms the savings bank account.

PROBLEMS

1. Suppose that you and your significant other each have a taxable income of $50,000 a year. The two of you could file your taxes as each being single, as being married filing jointly, or as being married filing separately (getting married is necessary for some of these choices, but I'm following the tax calculation thread here). Calculate your taxes in each situation. Is there a tax advantage to one of the situations?

2. Consider the same problem as above, but in this case, one of you has a taxable income of $10,000 while the other has a taxable income of $90,000.

3. This problem continues the study of Chapter 8, Comparing Loans. You have the choice of two competing mortgage loans, both for $350,000. You can get a 6% APR 15-year loan or a 6.5% APR 20-year loan. The interest on these loans will be deductible from your taxes. This year your taxable income will be $50,000. You have a very secure job, and you know that your taxable income will increase by 2% a year for the next 20 years, making the payments not an issue. Which loan is the better financial deal? Assume the loan is taken on January 1 and that the savings APR is 3.0%. Also, assume that the tax tables won't change and that I can use the 2008 married filing jointly table.

4. When you retire, you want to spend $40,000 a year from your savings to help support yourself. Inflation is running at 3% a year, however, and you notice that it costs you a little more each year to live the same way as you did the first year, so that you have to increase your initial $40,000 withdrawal each year. If you start out with $500,000 in your savings account, how many years will your money last? Assume that your savings are earning 5% interest a year.

5. You are (one of) a married couple filing jointly. Your taxable income this year will be about $100,000. You have about $25,000 to save or invest. Inflation is running at 2.5% annually. What APRs must you receive from a savings bank or from a tax-free investment to show an actual growth of 2% for the year? By actual growth, I mean growth in buying power.

Chapter 10

Life Insurance

10.1 WHAT IS AN INSURANCE POLICY?

An insurance policy is a contract, usually valid over a specified period of time or term, between you and an insurance company that basically reads the following:

1. You will give the insurance company some money. This money may be an up-front payment or it may be made up of periodic payments. The details will be spelled out in the insurance policy.

2. The insurance policy will describe some relatively unlikely occurrence during the term of the contract such as you breaking an elbow or having a fender bender or having your house robbed or dying and so on.

3. If the specified event occurs, the insurance company will pay you some money. The amount may be a specified fixed amount such as in a "$25,000 life insurance policy," or it might be an amount determined by the circumstances, such as the medical costs for setting your elbow or the cost of getting a new left front fender for your car.

The event that triggers the insurance company to pay you is an event that could not have been predicted, except in statistical generalities. This means that understanding insurance requires a bit of the understanding of mathematical probabilities.

Another name for buying an insurance policy is "placing a bet." Insurance companies don't dwell on this aspect, but buying an auto collision policy is essentially you saying, "I'll bet I'm going to have an auto accident next year," and the insurance company replying, "We'll take that bet. We don't think you're going to have an auto accident next year." Since having an auto accident is a relatively unlikely event, there are odds involved in the bet. For example, you might pay $1,000 for a policy that will cover up to $25,000 in auto repairs. What differentiates this kind of bet from a gambling casino bet is that when you buy an insurance policy, you don't want to win your bet. That is, you don't want to have an auto accident or to die.

Understanding the Mathematics of Personal Finance: An Introduction to Financial Literacy, by Lawrence N. Dworsky
Copyright © 2009 John Wiley & Sons, Inc.

An insurance policy that pays out if you die is called a life insurance policy. The amount you pay for it is called the premium.

10.2 PROBABILITY

Life insurance premiums are based on life insurance tables, which are issued annually by the U.S. federal government.[1] The basic idea is that nobody knows when they are going to die, but there's a lot of reliable information about the average life span of a large group of people like them. There are tables for the entire population, for all men, for all women, for men by race, for women by race, and so on. Table 10.1 shows the 2004 Life Table for all men. There's a lot of information in this table, much of it not useful to us right now. On the other hand, there's a lot of information that is useful to us. Before going through the table and explaining what the entries tell us, it's necessary to talk briefly about two topics: probability and expected, or average, value.

The mathematical field of probability (and its closely related field of statistical inference, or more simply, statistics) is fascinating and subtle. Defining some of the basic terms often takes chapters in textbooks. For our purposes, I'll offer some seat-of-the-pants definitions that should get us through what we need without detailed study but would not compel a mathematician to send this book through the garbage disposal.

Some events that we see every day are determined by so many subtle causes, most of which are always changing, that the event itself seems somewhat random. When you shuffle a deck of cards for example, the motion of each card is determined by where it started in the deck, how clean the surfaces of the cards are, the humidity that day, how stiffly you are holding the deck, and so on. All of these factors, and a host more that I've overlooked, contribute in such a complicated manner to the motion and ultimate position of each card in the deck that even the most cynical gambling houses from Las Vegas to Monte Carlo consider the locations of the individual cards in a well-shuffled deck to be random. When you turn over the top card of a shuffled deck, you don't know what to expect.

Of the 52 choices of cards you might turn over, there are 13 clubs, 13 spades, 13 hearts, and 13 diamonds (adding up to the 52 cards in a deck that has no jokers). If you repeat the basic experiment of shuffling the deck and turning over the top card many, many times, you would expect to see a diamond about one time out of four. That is, we say that the probability of turning over a diamond is equal to the number of opportunities of getting a diamond (13) divided by the total number of different cards that could appear (52). Doing the math, $13/52 = 1/4 = 0.25$.

The definition above is a simple but perfectly usable definition of probability: For a random event with D possible outcomes, if we're interested in N of them, then the probability of N occurring is N/D.

---

[1] The link http://www.cdc.gov/nchs/products/life-tables.htm will provide you with tables and a lot of information about how the tables are generated. The tables in this book reprinted from this document are courtesy of the National Center for Health Statistics, E. Arias, United States life tables, 2004. In *National Vital Statistics Reports*, vol. 56, no. 9 (Hyattsville, MD: National Center for Health Statistics, 2007).

Table 10.1   The 2004 U.S. Life Table for All Men

Age	$q(x)$	$l(x)$	$d(x)$	$L(x)$	$T(x)$	$e(x)$
0	0.007475	100,000	747	99,344	7,517,501	75.2
1	0.000508	99,253	50	99,227	7,418,157	74.7
2	0.000326	99,202	32	99,186	7,318,929	73.8
3	0.000250	99,170	25	99,157	7,219,744	72.8
4	0.000208	99,145	21	99,135	7,120,586	71.8
5	0.000191	99,124	19	99,115	7,021,451	70.8
6	0.000182	99,105	18	99,096	6,922,336	69.8
7	0.000171	99,087	17	99,079	6,823,240	68.9
8	0.000152	99,070	15	99,063	6,724,161	67.9
9	0.000125	99,055	12	99,049	6,625,098	66.9
10	0.000105	99,043	10	99,038	6,526,049	65.9
11	0.000111	99,033	11	99,027	6,427,011	64.9
12	0.000162	99,022	16	99,014	6,327,984	63.9
13	0.000274	99,006	27	98,992	6,228,970	62.9
14	0.000431	98,978	43	98,957	6,129,978	61.9
15	0.000608	98,936	60	98,906	6,031,021	61.0
16	0.000777	98,876	77	98,837	5,932,116	60.0
17	0.000935	98,799	92	98,753	5,833,278	59.0
18	0.001064	98,706	105	98,654	5,734,526	58.1
19	0.001166	98,601	115	98,544	5,635,872	57.2
20	0.001266	98,486	125	98,424	5,537,328	56.2
21	0.001360	98,362	134	98,295	5,438,904	55.3
22	0.001419	98,228	139	98,158	5,340,609	54.4
23	0.001435	98,089	141	98,018	5,242,451	53.4
24	0.001419	97,948	139	97,878	5,144,433	52.5
25	0.001390	97,809	136	97,741	5,046,554	51.6
26	0.001365	97,673	133	97,606	4,948,813	50.7
27	0.001344	97,540	131	97,474	4,851,207	49.7
28	0.001336	97,408	130	97,343	4,753,733	48.8
29	0.001341	97,278	130	97,213	4,656,390	47.9
30	0.001352	97,148	131	97,082	4,559,177	46.9
31	0.001371	97,017	133	96,950	4,462,094	46.0
32	0.001408	96,884	136	96,815	4,365,144	45.1
33	0.001469	96,747	142	96,676	4,268,329	44.1
34	0.001553	96,605	150	96,530	4,171,653	43.2
35	0.001653	96,455	159	96,375	4,075,123	42.2
36	0.001770	96,296	170	96,210	3,978,747	41.3
37	0.001911	96,125	184	96,033	3,882,537	40.4
38	0.002075	95,942	199	95,842	3,786,504	39.5
39	0.002254	95,742	216	95,635	3,690,662	38.5
40	0.002438	95,527	233	95,410	3,595,027	37.6
41	0.002632	95,294	251	95,168	3,499,617	36.7

Table 10.1   Continued

Age	q(x)	l(x)	d(x)	L(x)	T(x)	e(x)
42	0.002853	95,043	271	94,907	3,404,448	35.8
43	0.003113	94,772	295	94,624	3,309,541	34.9
44	0.003412	94,477	322	94,316	3,214,917	34.0
45	0.003735	94,154	352	93,979	3,120,601	33.1
46	0.004071	93,803	382	93,612	3,026,622	32.3
47	0.004428	93,421	414	93,214	2,933,010	31.4
48	0.004806	93,007	447	92,784	2,839,796	30.5
49	0.005206	92,560	482	92,319	2,747,012	29.7
50	0.005648	92,078	520	91,818	2,654,693	28.8
51	0.006121	91,558	560	91,278	2,562,875	28.0
52	0.006594	90,998	600	90,698	2,471,597	27.2
53	0.007045	90,398	637	90,079	2,380,899	26.3
54	0.007488	89,761	672	89,425	2,290,819	25.5
55	0.007946	89,089	708	88,735	2,201,394	24.7
56	0.008459	88,381	748	88,007	2,112,659	23.9
57	0.009064	87,633	794	87,236	2,024,652	23.1
58	0.009810	86,839	852	86,413	1,937,416	22.3
59	0.010706	85,987	921	85,527	1,851,002	21.5
60	0.011763	85,067	1,001	84,566	1,765,476	20.8
61	0.012934	84,066	1,087	83,522	1,680,909	20.0
62	0.014159	82,979	1,175	82,391	1,597,387	19.3
63	0.015362	81,804	1,257	81,175	1,514,996	18.5
64	0.016558	80,547	1,334	79,880	1,433,820	17.8
65	0.017847	79,213	1,414	78,507	1,353,940	17.1
66	0.019331	77,800	1,504	77,048	1,275,433	16.4
67	0.020992	76,296	1,602	75,495	1,198,386	15.7
68	0.022858	74,694	1,707	73,840	1,122,891	15.0
69	0.024921	72,987	1,819	72,077	1,049,050	14.4
70	0.027065	71,168	1,926	70,205	976,973	13.7
71	0.029363	69,242	2,033	68,225	906,768	13.1
72	0.032031	67,209	2,153	66,132	838,543	12.5
73	0.035178	65,056	2,289	63,912	772,411	11.9
74	0.038734	62,767	2,431	61,552	708,499	11.3
75	0.042414	60,336	2,559	59,057	646,947	10.7
76	0.046171	57,777	2,668	56,443	587,891	10.2
77	0.050325	55,109	2,773	53,723	531,448	9.6
78	0.055085	52,336	2,883	50,894	477,725	9.1
79	0.060498	49,453	2,992	47,957	426,831	8.6
80	0.066557	46,461	3,092	44,915	378,873	8.2
81	0.072986	43,369	3,165	41,786	333,958	7.7
82	0.079682	40,204	3,204	38,602	292,172	7.3
83	0.086593	37,000	3,204	35,398	253,570	6.9

Table 10.1   Continued

Age	q(x)	l(x)	d(x)	L(x)	T(x)	e(x)
84	0.094013	33,796	3,177	32,207	218,172	6.5
85	0.102498	30,619	3,138	29,050	185,965	6.1
86	0.111640	27,481	3,068	25,947	156,915	5.7
87	0.121472	24,413	2,965	22,930	130,968	5.4
88	0.132023	21,447	2,832	20,031	108,039	5.0
89	0.143319	18,616	2,668	17,282	88,007	4.7
90	0.155383	15,948	2,478	14,709	70,726	4.4
91	0.168232	13,470	2,266	12,337	56,017	4.2
92	0.181880	11,204	2,038	10,185	43,680	3.9
93	0.196334	9,166	1,800	8,266	33,496	3.7
94	0.211592	7,366	1,559	6,587	25,229	3.4
95	0.227645	5,808	1,322	5,147	18,642	3.2
96	0.244476	4,486	1,097	3,937	13,496	3.0
97	0.262057	3,389	888	2,945	9,559	2.8
98	0.280351	2,501	701	2,150	6,614	2.6
99	0.299312	1,800	539	1,530	4,463	2.5
100 or over	1.00000	1,261	1,261	2,933	2,933	2.3

If we flip a coin, there are two possible outcomes: heads or tails. The probability of heads is therefore $1/2 = 0.5$, as is the probability of tails. The total of all the probabilities (a long way of saying that one of the possible choices *must* occur) is always 1.0.

Extending this a bit, the probability of getting 5 heads when I flip 10 coins (or equivalently, flip one coin 10 times) is $5/10 = 0.5$. If you try this, however, don't be surprised if you don't get exactly 5 heads out of 10 flips. What happens in practice is that the larger the number of times you try, the more likely it is that the percentage of error (from exactly 50% heads) gets smaller.

Let's take this basic idea and see how this works with a hypothetical life insurance company. Suppose that you and your friends get together and decide to all put some money into a bank for life insurance for all of you. I'm setting the situation up this way so that I can avoid issues of company profits, operating expenses, and so on. This is an idealized group—there are no profits, no taxes, no cost of doing business, and so on.

Imagine that the life insurance tables tell you (and I'll show you soon how they tell you this) that the probability of each of the people in your group dying within the next year is 0.02 (= 1/50). This means that if there are 100 of you, it's most probable that two of you will die during the next year. If you each put $1,000 into the bank account, that's a total of $100,000. If two of you die, there's funding to give the families of these two people $50,000 each. Note that in this ideal situation, there is no money left in the bank account at the end of the year and you have to start over again next year. In the jargon of life insurance, you've each paid a $1,000 premium for a 1-year, $50,000 term insurance policy.

This scenario works fine unless more than two people die during the year. If, say, four people die, then there's only enough money to pay $25,000 to each of the decedents' families. This is an example of where betting on the probabilities can get you into trouble. When you have a low probability of something happening and a small enough group for this thing to happen to, the actual number of occurrences of your event of interest can wander all over the place.

We don't want to resolve this dilemma by somehow making it more likely for people to die during the next year. Instead, we avoid 100 people insurance "groups" and try to deal with very large numbers of people. For example, if we had 100,000 people, each contributing $1,000, we would collect $10,000,000 dollars into our insurance pool. Using the same probability of dying, 2% of 100,000 is 2,000 and we'd still be able to give $50,000 to the family of each person who died.

With our small group, if we wanted to collect enough money to pay for more people than the ideal probability tells us we'll have to pay for, we have to collect another $500, or a total of $1,500 from each person. With our large group, we'd only have to collect another $2 from each person to pay for each additional person. We could cover a whole bunch of extra people.

What if we still haven't put aside enough money for the actual number of people who die? Insurance companies handle this problem by noting that averages tend to be averages. That is, if one of the policy groups (e.g., 50-year-old men) requires more funding than was predicted for this year, there will be another group (e.g., 60-year-old women) that will require less funding than was predicted this year. When all else fails, insurance companies can cross-insure each other.

Calculating how much extra money should be available "just in case" involves discussions of confidence intervals that are, in a sense, probabilities of probabilities. Such discussion is beyond the scope of this book. It isn't really that hard to understand; it's just a lengthy discussion that heads off in a direction that's not central to this book. Figures 10.1 and 10.2 show the results of one of these calculations for the examples presented above.

In Figure 10.1, the horizontal axis is the number of people who will die during the year in our 100-person group, and the vertical axis is the relative probability of this number of people dying.[2] As can be seen, the maximum probability is for two people to die. Interestingly, it is almost as likely (i.e., as probable) for only one person do die. Of more importance, however, is that it is about one-third as likely for four people to die as it is for two people to die and about one-tenth as likely for five people to die. In other words, it's not unlikely that twice or even more than twice the number of people will die than our basic funding model provides for.

---

[2] Relative probability means that the probability data are being presented without numbers on the vertical axis. My reason for doing this is that the numbers would not contribute to the discussion but would take a long time to explain. The intent is just to be able to compare different values of probability on the graph, that is, *relative probability*.

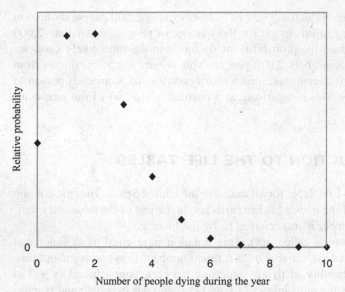

Figure 10.1   Probable number of deaths in a 100-person group with 2% probability of dying.

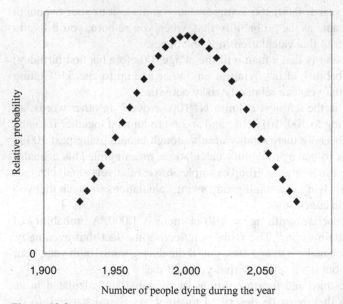

Figure 10.2   Probable number of deaths in a 100,000-person group with 2% probability of dying.

Now look at Figure 10.2. In this case, the horizontal axis is the number of people who will die during the year in our 100,000-person group. The most likely case is for 2,000 people to die (2% of 100,000). It's almost as likely that 2,001 or 2,002 or 1,999 or 1,998 people will die as it is that exactly 2,000 people to die. This is another way of saying that there's "nothing special" about exactly 2,000 people dying. On the other hand, looking at this graph for the number of people greater than 2,000 that have about one-third the probability of dying as in the most likely case, we see that this probability is about 2,070 people. This is only a 3.5% deviation from the most likely case. It doesn't take much extra contribution from each person to cover this contingency. This comparison, in a nutshell, shows why insurance companies work.

10.3 INTRODUCTION TO THE LIFE TABLES

Table 10.1 is the 2004 Life Table for all males in the United States. The first column to the left is the age of the man at his last birthday. In the rest of the table entry definitions, this single number is just referred to by the letter x.

The second column is called $q(x)$. This notation is referred to as functional notation. It's saying that the variable q "is a function of x." Don't worry about this. Column q is the probability of dying sometime between your xth and $(x + 1)$st birthday. For example, the probability of dying between your twentieth and twenty-first birthday, that is, when you're 20 years old, is 0.001266. This definition, as stated, is correct—but should be elaborated on. This probability that you'll die during your twentieth year is 0.001266 assumes that you made it to your twentieth birthday. It is *not* the same as the probability that, when you're born, you'll live to be 20 years old. I'll get to that calculation in a few pages.

Note that the probability that a man will die at age 0 (before his first birthday) is higher than the probability of his dying at any other age up to age 54. Getting born and living your first year are relatively risky activities.

The bottom entry in the leftmost column is "100 or over." In other words, all the probabilities of living to 100, 101, 102, and so on are lumped together into one catchall entry. This is because there simply aren't enough people living past 100 in the United States today to make probability calculations meaningful. This is analogous to the 100-person life insurance group example above; relatively small changes in the number of people dying in a small group swing calculations so much that you can't draw very reliable conclusions.

The value of q associated with age = 100 or more is 1.000. A probability of 1.00 is called the "certain event." The table is reflecting the fact that eventually, everyone will die. If you make it to age 100, you'll die during your 100th year, your 101st year, and so on, but it's a certainty that you will die.

The entries in the third and fourth columns of the table are calculated in an interactive manner, so they must be described together. As I've discussed, to do meaningful probability calculations on a relatively unlikely event (an event that has a probability much less than 1.00), we need a large group of people. It is conventional to use a group of 100,000 as a "standard starting point." I don't know if this is the best size to start with, or even how to calculate if it's the best size to start with, but it is the standard starting point, so that's where this table starts the third column: for $x = 0$ (age = 0–1), $l = 100,000$.

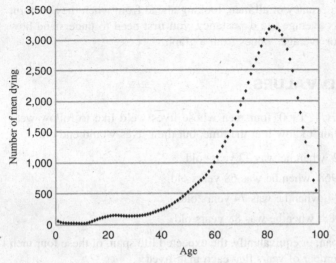

Figure 10.3   Probable of number of men dying each year after birth out of the original 100,000 group based on the 2004 Life Table.

If there are 100,000 men starting out and the probability of each of them dying during their first year of life is 0.007475, then during this year (0.007475) (100,000) = 747 men will die.[3]

If I start with 100,000 men and during the year 747 of them die, then going into the next year of life ($x = 1$), there are 100,000 – 747 = 99,253 men alive. This is the second entry in the l column.

Now the work gets repetitive: Looking at the $x = 1$ line in the table, there are 99,253 men starting out the year, and there's a probability of 0.000508 that each of them will die. The number of men most likely to die during this year is therefore (99,253)(0.000508) = 50. This is the second entry in the d column.

Going to the third entry of column l, 99,253 men started the year and 50 of them died during the year. The column l entry for the $x = 2$ line is therefore 99,253 – 50 = 99,202 (again, you can't see the rounding that actually occurred).

Following this pattern all the way down to the $x = 100$ or more line, the column l entry is 1,261 men reaching their one hundredth birthday. Since the probability of each of them dying during their one hundredth year or any year thereafter is 1.000, the column d entry is (1,261)(1.00) = 1,261. At this point, there's nobody left of the original 100,000 men and we stop.

Figure 10.3 is a graph of column d. This graph peaks at about age 83. This means that the most likely age for a man to die is 83 years old. It does *not* mean

---

[3] Looking at the table as shown, the multiplication yields 747.5, which is properly rounded to 748. This is because I chose to show the table to only six figures after the decimal point. Had I shown it to seven figures, you would have seen that the probability is really 0.0074748. The multiplication now yields 747.48 that, to the nearest full person, is 747. Careless rounding of numbers can lead to misleading results, and you could go as far as creating criminal schemes that might pass unnoticed by deliberately engaging in this type of activity.

that the average life expectancy of all men, looking ahead from when they're born, is 83 years. To find this average life expectancy, you first need to understand how to calculate expected, or average, values from a graph.

10.4 EXPECTED VALUES

Assume that on January 1, 1900, four men whose lives we'd like to follow were born. They of course didn't know it at the time, but their lives would end on:

First man: In 1922, when he was 22 years old.

Second man: In 1968, when he was 68 years old.

Third man: In 1974, when he was 74 years old.

Fourth man: In 1988, when he was 88 years old.

The average life span, or equivalently, the expected life span, of these four men is the average of the number of years that each man lived:

$$\frac{22 + 68 + 74 + 88}{4} = 63.0.$$

Suppose we started studying these men in 1950. In 1950, there are only three of these men left alive. If we kept track of them until they were all dead, we would find that their expected life span is

$$\frac{68 + 74 + 88}{3} = 76.7.$$

Many people interpret results such as these as telling them that because they have lived some number of years (in this case, 50), their life expectancy is higher than it was when they were born. This interpretation is nonsense. Their life expectancy hasn't changed. What has changed is that the people who, for whatever reason, were destined to die earlier have already died and are no longer being counted in the averaging process. You can see this clearly in the two average calculations above. The second calculation is the average of the life spans of the three men who hadn't died before they were 22 years old.

Column d of the Life Table is a list of life spans just like the one used in the examples above. To find the expected life span of all men from the time they're born, we just add up all 100,000 life spans in the numerator, put the number 100,000 in the denominator, and perform the calculation.

This procedure will work, but it certainly is a mess to set up, even on a computer spreadsheet. Fortunately, there are more efficient ways to perform this calculation. If you're interested, look up the topic "weighted averages" on the Web or in a text. The important point to remember is that when all is said and done, the calculation is exactly the same as the two calculations I did above.

If I do perform this calculation, I find that the expected life span for all men, at birth, is 74.7 years.

What is the expected life span starting at, say, 10 years? To calculate this, do the same thing I did in the simple example. Repeat the first calculation but eliminate everybody who died before their tenth birthday. The answer comes out to be 75.4 years.

Column e of the Life Table shows this same calculation but is expressed a bit differently. Using the line $x = 10–11$ years (10 years old), first subtract 10 from 75.4 so that we're talking about the number of years left to live rather than the life span. Then, add half a year so as to get a midyear value—this is more representative of people between 10 and 11 years old than a calculation taken just at the tenth birthday. Putting all of this together,

$$e = 75.4 - 10 + 0.5 = 65.9.$$

Going the other way, if you want to know the average life span of a 50-year-old man, start with the value of e in the table. For $x = 50–51$, $e = 28.8$. Subtract 0.5 to get the reference back to the beginning of the year: $28.8 - 0.5 = 28.3$. Then, add 50, resulting in $28.3 + 50 = 78.3$ as the expected life span for 50-year-old men.

Figure 10.4 shows the expected life span versus the current age. There are two sets of data shown: the expected life spans for men and for women. Women, on the average, live longer than men. This will be an important consideration when we consider reverse mortgages in Chapter 12 because the insurance companies will have to worry about who is the second spouse to die. For now, it's just an interesting fact.

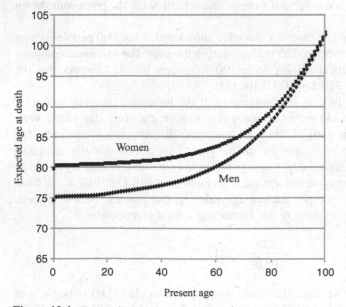

Figure 10.4 Expected age at death for men and women versus present age.

10.5 TERM INSURANCE

The simplest kind of life insurance policy you can buy is term insurance. As the name implies, a term insurance policy insures you for a specific term. You pay a premium up front for the policy. If you don't die during the term, you and the insurance company no longer have any contractual relationship. Whatever happens, the insurance company keeps the premium. Actually, term insurance is the only kind of life insurance there is. All the other life insurance products that you can buy are built on term insurance policies, sometimes combined with savings/investment accounts.

Term insurance policies themselves can get very complicated. They might be written for many years, and payments made might be spread out over these years rather than up front. There are so many possible variations that it's impossible to cover them all. But let's look at a few simple cases. I will be losing accuracy by considering only average values and by not considering the business costs, the return on investment of all the money that the insurance company is holding, the need for the insurance company to make a profit for its owners (investors), and so on. Also, life insurance companies have many sets of life tables that get far more specific than the few sets I'm using for my examples. In other words, I will show you what goes on in putting together a life insurance policy. I can't actually price your policy.

Suppose that a 50-year-old woman wants a 1-year, $100,000 term life insurance policy for the following year. Table 10.2 shows the 2004 Life Table for all women. This table just shows the age (x) and the probability of dying at that age (q) because that's all that's needed.

Table 10.3 shows a part of Table 10.2 from age 50 to age 100. In addition, I have added column l and column d, with column l starting at 100,000 people. The numbers for column l and column d were generated following the procedure shown above.

Looking at the top line (age 50), the table shows that if 100,000 people sign up for this policy, on the average, 320 will die during the year. The insurance company will have to pay out 320($100,000). Since 100,000 people bought these policies, the cost per person is just 320($100,000)/100,000 = $320.

Let's make this a little more complicated. If the insurance company sold a lot of these policies, it would expect to see a few women die every day of the year, since some women die early in the year, and some die very late in the year. The average date of death is the middle of the year. This means that the insurance company gets to hold everybody's money for an average time of half a year. If an average woman has to pay an amount today that will be worth $320 half a year from now, then she should only pay the present value of that amount. At 4% interest, therefore, she should only have to pay (assuming annual compounding)

$$\frac{\$320}{(1.04)^{0.5}} = \$313.80.$$

What if this same woman, who's just turning 50 years old, wants to buy a 2- or more year term policy? At first blush, you might ask why she would do this. If she

Table 10.2   The 2004 U.S. Life Table for All Women (Age and q Columns Only)

Age	q	Age	q	Age	q
0–1	0.006091	34–35	0.000825	68–69	0.014966
1–2	0.000457	35–36	0.000892	69–70	0.016407
2–3	0.000267	36–37	0.000971	70–71	0.017945
3–4	0.000197	37–38	0.001071	71–72	0.019617
4–5	0.000168	38–39	0.001190	72–73	0.021503
5–6	0.000151	39–40	0.001321	73–74	0.023635
6–7	0.000138	40–41	0.001453	74–75	0.025987
7–8	0.000129	41–42	0.001586	75–76	0.028358
8–9	0.000120	42–43	0.001727	76–77	0.030849
9–10	0.000112	43–44	0.001883	77–78	0.033818
10–11	0.000107	44–45	0.002055	78–79	0.037481
11–12	0.000113	45–46	0.002243	79–80	0.041792
12–13	0.000135	46–47	0.002439	80–81	0.046463
13–14	0.000178	47–48	0.002633	81–82	0.051306
14–15	0.000237	48–49	0.002819	82–83	0.056613
15–16	0.000306	49–50	0.003005	83–84	0.062608
16–17	0.000371	50–51	0.003204	84–85	0.069533
17–18	0.000421	51–52	0.003432	85–86	0.076645
18–19	0.000446	52–53	0.003695	86–87	0.084411
19–20	0.000453	53–54	0.004000	87–88	0.092876
20–21	0.000456	54–55	0.004346	88–89	0.102085
21–22	0.000464	55–56	0.004725	89–90	0.112081
22–23	0.000471	56–57	0.005137	90–91	0.122907
23–24	0.000481	57–58	0.005594	91–92	0.134602
24–25	0.000492	58–59	0.006110	92–93	0.147201
25–26	0.000506	59–60	0.006697	93–94	0.160735
26–27	0.000522	60–61	0.007389	94–95	0.175225
27–28	0.000541	61–62	0.008167	95–96	0.190689
28–29	0.000565	62–63	0.008977	96–97	0.207132
29–30	0.000593	63–64	0.009776	97–98	0.224550
30–31	0.000627	64–65	0.010581	98–99	0.242924
31–32	0.000667	65–66	0.011466	99–100	0.262224
32–33	0.000712	66–67	0.012498	100 or over	1.00000
33–34	0.000764	67–68	0.013661		

just planned to buy term policies year by year and then died, say, during the first year, her premiums for the subsequent years would still be part of her estate rather than in the hands of the insurance company. The need to buy such a policy can arise, for example, if there is a business loan with the repayment due as a lump sum, say, 5 years from today. The creditor might want to guarantee his or her repayment in case the borrower dies before the repayment is due, without having to get involved

Table 10.3 Life Table for Women Aged 50 and Up

x	q	l	d	x	q	l	d
50–51	0.003204	100,000	320	76–77	0.030849	74,058	2,285
51–52	0.003432	99,680	342	77–78	0.033818	71,774	2,427
52–53	0.003695	99,337	367	78–79	0.037481	69,347	2,599
53–54	0.004000	98,970	396	79–80	0.041792	66,747	2,790
54–55	0.004346	98,574	428	80–81	0.046463	63,958	2,972
55–56	0.004725	98,146	464	81–82	0.051306	60,986	3,129
56–57	0.005137	97,682	502	82–83	0.056613	57,857	3,275
57–58	0.005594	97,181	544	83–84	0.062608	54,582	3,417
58–59	0.006110	96,637	590	84–85	0.069533	51,165	3,558
59–60	0.006697	96,046	643	85–86	0.076645	47,607	3,649
60–61	0.007389	95,403	705	86–87	0.084411	43,958	3,711
61–62	0.008167	94,698	773	87–88	0.092876	40,247	3,738
62–63	0.008977	93,925	843	88–89	0.102085	36,509	3,727
63–64	0.009776	93,082	910	89–90	0.112081	32,782	3,674
64–65	0.010581	92,172	975	90–91	0.122907	29,108	3,578
65–66	0.011466	91,197	1,046	91–92	0.134602	25,530	3,436
66–67	0.012498	90,151	1,127	92–93	0.147201	22,094	3,252
67–68	0.013661	89,024	1,216	93–94	0.160735	18,842	3,029
68–69	0.014966	87,808	1,314	94–95	0.175225	15,813	2,771
69–70	0.016407	86,494	1,419	95–96	0.190689	13,042	2,487
70–71	0.017945	85,075	1,527	96–97	0.207132	10,555	2,186
71–72	0.019617	83,548	1,639	97–98	0.224550	8,369	1,879
72–73	0.021503	81,909	1,761	98–99	0.242924	6,490	1,577
73–74	0.023635	80,148	1,894	99–100	0.262224	4,913	1,288
74–75	0.025987	78,254	2,034	100 or over	1.00000	3,625	3,625
75–76	0.028358	76,220	2,161				

in chasing her estate for his or her money. If there is a life insurance policy with the creditor as beneficiary, then he or she doesn't have to worry about getting his or her money back if she dies before the loan is due. The creditor would demand a paid-up term life insurance policy at the start of the loan period (when the woman gets the money), and the lender would have to consider the premium for this policy as part of her cost of getting the loan (i.e., increasing the effective interest rate).

Returning to Table 10.3, look at the fifty-first year. There are $100,000 - 320 = 99,680$ women starting their fifty-first year, so we should expect $0.003432(99,680) = 342$ deaths during this year.

During the first year of the 2-year policy, the insurance company has to pay out 320 times for every 100,000 people that signed up, as discussed above. In the second year, it has to pay out 342 times for every 100,000 people that originally signed up.

I just showed how much the 1-year term policy would be for the 50-year-old woman. This is identical to the cost of the first year of a 2-year term policy.

For the second year, the cost to the insurance company for a $100,000 policy, following the same procedure, is $342 per person in the original (100,000-person) group.

This cost is incurred by the insurance company, on the average, 1.5 years after it collected the premiums on the policies, and the present value of this cost at the date of collecting the premiums is

$$\frac{\$342}{(1.04)^{1.5}} = \$332.46.$$

The premium for the 2-year policy is the sum of the two individual premiums, $313.80 + $332.46 = $646.26.

This procedure may be continued for as many years as you want the policy.

10.6   TIME PAYMENTS

If you want to buy a 20-year policy using the data shown in Table 10.3, the premium is $9,340. What if you don't have this money available? The reasonable answer is to take a loan. You'll be insured for 20 years; why not "pay as you go?"

A 20-year loan for $9,340, at 6% compounded monthly, requires payments of about $67 each month. This doesn't seem like a lot but keep in mind that this is just an example.

If you take the loan, it looks like everything is taken care of. The insurance company has its premium, you have your 20-year insurance policy, and you have monthly payments on the loan you took that you can handle. There is, however, one more item to consider here—what happens if you do die during the 20-year period? The insurance company pays $10,000 to your beneficiary, as was promised. However, your estate still owes the balance on your loan.

If you die exactly 10 years after you took the policy out, on your sixtieth birthday, then there is a loan balance of $6,030. One way to balance the books is for the beneficiary of the insurance policy to pay off the loan. This means that, in effect, the insurance policy only paid the beneficiary $100,000 – $6,030 = $93,970.

If, for whatever reason, you wanted your beneficiary to receive the full intended amount, then you could purchased a $110,000 policy. The premium for this policy and the monthly payments on your loan would both go up to 10%.

Another way to handle the time payment situation is for the insurance company to be the banker. That is, the insurance company, acting internally like a bank, arranges to pay your premium up front and collect monthly premiums from you. This has the advantage that the insurance company can look at the Life Tables and come up with an average balance left on the loan after many people in your situation arrange this package deal and then calculate how much insurance it must sell you (internally, on its books) so that, on the average, there is enough money to pay the beneficiary exactly $100,000 while repaying the balance of the loan. This arrangement also has the advantage of simplicity. When you die, your estate might be split between heirs who are not necessarily identically the beneficiary(ies) of your life insurance.

10.7   DECREASING TERM INSURANCE

Another type of term life insurance is decreasing term life insurance. A decreasing term life insurance policy pays a little less each year for the term of the insurance. As an example, consider a $100,000, 20-year decreasing term policy. In year 1, the policy pays $100,000. In year 2, it pays $95,000 and so on, up to year 20, when it would pay $5,000.

At first blush, this seems like an odd sort of life insurance. Why would you want a beneficiary to receive less if you die later? This type of insurance is the equivalent of the loan insurance described in the last section, but it is more appropriate when it's insuring a loan that is being paid off over time. If you are paying a loan back with regular payments, then each month you owe less. Consequently, you can'save on your insurance premium if you buy insurance that pays less as time goes buy.

Decreasing term life insurance is often sold under the name of mortgage insurance. This is a policy that pays off your mortgage if you die so that your family doesn't lose your house. Auto loan providers often offer similar loans. In both of these cases, the loans might be a combination of life insurance and some sort of health and/or disability insurance. That is, in addition to paying off your loan if you die, the policy will make payments on the loan if you're sick and possibly pay off the loan if you're disabled and can no longer earn a living. Since there are so many possible variations of this combination of insurance policies, it is very important that this type of policy be scrutinized carefully so that you know what you're buying. Typically, these policies are fully paid for up front, but I guess that anything is possible.

Table 10.4 shows the year-by-year calculations, for the same person as above, for a 20-year, $100,000 decreasing term policy that decreases by $5,000 per year. This table could have been built from scratch using the same process that I used for Table 10.3, but I took a shortcut; I started with Table 10.3. Then, I simply scaled the costs each year. That is, the costs for the second year (a $95,000 policy) are just 19/20 (= 0.95) of the original costs for that year. I referred to this multiplication, or scaling, factor as scale in the table. The cost for the third year (a $90,000 policy) is just 18/20 (= 0.90) of the original costs for that year and so on. In the table, I showed the numbers for the $100,000, 20-year term policy as well as the numbers for the decreasing term policy. This should make it a bit easier to follow where the decreasing term policy numbers came from.

The premium for the policy, again, is the sum of all the present value costs of the right-hand column, which is $4,278. This is considerably less expensive than the 20-year term life insurance policy for $9,340, as you can see in the table.

10.8   INSURANCE FOR THE REST OF YOUR LIFE

What if you want a life insurance not for a specific period such as 20 years but simply for the rest of your life? This type of policy would be different from that which I've already presented. In the previous examples, the insurance company never knew if it would have to pay. The insured person could still be alive at the

Table 10.4 A 20-Year Decreasing Term $100,000 Life Insurance Policy

Age	q	l	d	Term PV ($)	Scale	Decreasing term PV ($)
50	0.003204	100,000	320	314.23	1.00	314.23
51	0.003432	99,680	342	322.59	0.95	306.46
52	0.003695	99,337	367	332.78	0.90	299.51
53	0.004000	98,970	396	345.14	0.85	293.37
54	0.004346	98,574	428	359.11	0.80	287.29
55	0.004725	98,146	464	373.76	0.75	280.32
56	0.005137	97,682	502	388.84	0.70	272.19
57	0.005594	97,181	544	405.06	0.65	263.29
58	0.006110	96,637	590	423.05	0.60	253.83
59	0.006697	96,046	643	443.17	0.55	243.74
60	0.007389	95,403	705	466.97	0.50	233.49
61	0.008167	94,698	773	492.62	0.45	221.68
62	0.008977	93,925	843	516.42	0.40	206.57
63	0.009776	93,082	910	535.87	0.35	187.56
64	0.010581	92,172	975	552.24	0.30	165.67
65	0.011466	91,197	1,046	569.35	0.25	142.34
66	0.012498	90,151	1,127	589.87	0.20	117.97
67	0.013661	89,024	1,216	612.21	0.15	91.83
68	0.014966	87,808	1,314	636.09	0.10	63.61
69	0.016407	86,494	1,419	660.50	0.05	33.03
		Total:		9,340		4,278

end of the term. With what you're asking for now, the insurance company, sooner or later, must pay.

Working this problem out is not difficult—we do the same thing we did in the previous examples, but we extend the upper age to 100. Since $q_{100} = 1$, the fact that the insurance company sooner or later has to pay is automatically included in the calculation. Without showing the details here, the policy for a 50-year-old woman comes to approximately $31,000. This is considerably more than the $9,340 it took to insure her from age 50 to age 69, but remember that we've not only more than doubled the policy term, but we're adding the older years where death is more probable, up to age 100 where, at least according to our tables, death is inevitable.

Figure 10.5 shows the premium for this kind of policy, for a man and for a woman, based on the age that he or she buys the policy. As you would expect, the women's policy is less expensive. Near age 100, either policy costs about $100,000. This is to be expected; if you buy a $100,000 policy close to age 100 and (according to the tables) you're not going to live past 100, you should be prepared to pay about what you're going to get back.

Some term policies are sold by the year. That is, at the beginning of each year the premium cost is based on the q value in the Life Table. For a man of age 20, for

Figure 10.5   Cost of a $100,000 lifetime policy versus age at purchase.

example, $q = 0.001266$, and therefore, before looking at costs and profits, the insurance company must charge $100,000 \times 0.001266 = \$127$. This doesn't seem like a lot. However, at age 70, that same $100,000 policy would cost $100,000 \times 0.02765 = \$2,707$. This adds up for a several hundred thousand dollar policy.

10.9   WHOLE LIFE INSURANCE

A popular policy with a fixed monthly payment is the so-called whole life policy. There are so many variations on how this can be structured that I can't possibly cover them all. I'll look at a 20-year-old typically healthy man and present one possible scenario. You should get the idea of what's going on pretty easily and be able to understand different variations when they're put before you.

As was shown above, for a 20-year-old man, a year's $100,000 1-year term insurance should only cost $127. Suppose, however, that this man pays $900 a year instead. I'll treat things annually even though payments are usually made monthly, just to keep the table showing these calculations from becoming uncomfortably long.

At the inception of the policy (our young man's twentieth birthday), he writes a check for $900. A total of $127 goes to buy him a $100,000 term life insurance policy for the year. The remaining $773 goes into a savings/investment account. If we assume that this account earns 4% a year, then on his twenty-first birthday, he has $(1.04)(\$473) = \$804$. On this day, he makes another payment (of $900), and the insurance company takes the required payment for the year's term insurance ($136) out of the account, leaving the account with a balance of $1,568. As we continue this procedure, year after year, the balance in the account grows—even though the annual 1-year term policy payment is also growing (Table 10.5). Note that I didn't

Table 10.5   Whole Life Insurance Policy Example

Age	q	l	d	Premium ($)	Payment ($)	Balance ($)
20	0.001266	100,000	127	126.60	900.00	773.40
21	0.001360	99,873	136	135.85	900.00	1,568.49
22	0.001419	99,738	142	141.52	900.00	2,389.71
23	0.001435	99,596	143	142.90	900.00	3,242.39
24	0.001419	99,453	141	141.08	900.00	4,131.00
25	0.001390	99,312	138	138.04	900.00	5,058.20
26	0.001365	99,174	135	135.39	900.00	6,025.14
27	0.001344	99,039	133	133.14	900.00	7,033.01
28	0.001336	98,905	132	132.13	900.00	8,082.20
29	0.001341	98,773	132	132.44	900.00	9,173.04
30	0.001352	98,641	133	133.35	900.00	10,306.62
31	0.001371	98,508	135	135.02	900.00	11,483.86
32	0.001408	98,373	139	138.52	900.00	12,704.70
33	0.001469	98,234	144	144.28	900.00	13,968.61
34	0.001553	98,090	152	152.29	900.00	15,275.06
35	0.001653	97,937	162	161.86	900.00	16,624.20
36	0.001770	97,776	173	173.04	900.00	18,016.13
37	0.001911	97,603	187	186.54	900.00	19,450.24
38	0.002075	97,416	202	202.11	900.00	20,926.13
39	0.002254	97,214	219	219.11	900.00	22,444.06
40	0.002438	96,995	236	236.44	900.00	24,005.39
41	0.002632	96,758	255	254.65	900.00	25,610.95
42	0.002853	96,504	275	275.33	900.00	27,260.06
43	0.003113	96,228	300	299.60	900.00	28,950.86
44	0.003412	95,929	327	327.28	0.00	29,781.62
45	0.003735	95,601	357	357.03	0.00	30,615.86
46	0.004071	95,244	388	387.73	0.00	31,452.77
47	0.004428	94,857	420	420.06	0.00	32,290.82
48	0.004806	94,437	454	453.89	0.00	33,128.57
49	0.005206	93,983	489	489.31	0.00	33,964.40
50	0.005648	93,493	528	528.01	0.00	34,794.97
51	0.006121	92,965	569	569.05	0.00	35,617.71
52	0.006594	92,396	609	609.22	0.00	36,433.20
53	0.007045	91,787	647	646.64	0.00	37,243.89
54	0.007488	91,141	682	682.49	0.00	38,051.15
55	0.007946	90,458	719	718.75	0.00	38,854.45
56	0.008459	89,739	759	759.10	0.00	39,649.54
57	0.009064	88,980	807	806.55	0.00	40,428.97
58	0.009810	88,174	865	864.95	0.00	41,181.18
59	0.010706	87,309	935	934.75	0.00	41,893.67
60	0.011763	86,374	1,016	1,016.03	0.00	42,553.40

Table 10.5 Continued

Age	q	l	d	Premium ($)	Payment ($)	Balance ($)
61	0.012934	85,358	1,104	1,104.01	0.00	43,151.53
62	0.014159	84,254	1,193	1,192.98	0.00	43,684.61
63	0.015362	83,061	1,276	1,276.01	0.00	44,155.98
64	0.016558	81,785	1,354	1,354.16	0.00	44,568.06
65	0.017847	80,431	1,435	1,435.42	0.00	44,915.36
66	0.019331	78,995	1,527	1,527.05	0.00	45,184.93
67	0.020992	77,468	1,626	1,626.23	0.00	45,366.10
68	0.022858	75,842	1,734	1,733.58	0.00	45,447.16
69	0.024921	74,109	1,847	1,846.84	0.00	45,418.21
70	0.027065	72,262	1,956	1,955.77	0.00	45,279.17
71	0.029363	70,306	2,064	2,064.39	0.00	45,025.94
72	0.032031	68,242	2,186	2,185.85	0.00	44,641.13
73	0.035178	66,056	2,324	2,323.71	0.00	44,103.06
74	0.038734	63,732	2,469	2,468.61	0.00	43,398.57
75	0.042414	61,263	2,598	2,598.44	0.00	42,536.08
76	0.046171	58,665	2,709	2,708.61	0.00	41,528.91
77	0.050325	55,956	2,816	2,816.02	0.00	40,374.05
78	0.055085	53,140	2,927	2,927.22	0.00	39,061.79
79	0.060498	50,213	3,038	3,037.80	0.00	37,586.46
80	0.066557	47,175	3,140	3,139.83	0.00	35,950.09
81	0.072986	44,035	3,214	3,213.97	0.00	34,174.13
82	0.079682	40,821	3,253	3,252.73	0.00	32,288.36
83	0.086593	37,569	3,253	3,253.19	0.00	30,326.70
84	0.094013	34,316	3,226	3,226.10	0.00	28,313.67
85	0.102498	31,089	3,187	3,186.59	0.00	26,259.62
86	0.111640	27,903	3,115	3,115.08	0.00	24,194.93
87	0.121472	24,788	3,011	3,011.03	0.00	22,151.70
88	0.132023	21,777	2,875	2,875.03	0.00	20,162.73
89	0.143319	18,902	2,709	2,708.97	0.00	18,260.27
90	0.155383	16,193	2,516	2,516.07	0.00	16,474.61
91	0.168232	13,677	2,301	2,300.85	0.00	14,832.74
92	0.181880	11,376	2,069	2,069.04	0.00	13,357.01
93	0.196334	9,307	1,827	1,827.23	0.00	12,064.06
94	0.211592	7,480	1,583	1,582.61	0.00	10,964.02
95	0.227645	5,897	1,342	1,342.40	0.00	10,060.17
96	0.244476	4,555	1,113	1,113.47	0.00	9,349.11
97	0.262057	3,441	902	901.75	0.00	8,821.33
98	0.280351	2,539	712	711.90	0.00	8,462.28
99	0.299312	1,827	547	546.96	0.00	8,253.81
100	1.00000	1,280	1,280	1,280.44	0.00	7,303.52

have to do any present value corrections to the premiums because they are being taken from the savings account on the day they are due.

After the annual payment at age 43, the policy is declared "paid off," and no further $900 payments are needed. The balance in the account still continues to grow because the annual interest outweighs the insurance policy payment, until age 68, when the term insurance premiums get large enough to outweigh the accruing interest. At age 100, if our young man lives that long, there is a balance in the account of approximately $7,300.

At the time of death, whenever that may be, the insurance company pays the beneficiaries the $100,000 face value of the policy and also the balance in the account.

This type of policy is attractive to many people for several reasons:

1. At age 47, the policy is "paid off" and no further premium payments are required.

2. Upon death, the beneficiaries get both the value of the life insurance policy and the balance of the account, the latter possibly being quite substantial. For example, if our young man dies at age 65, there is almost $65,000 in the account.

3. This policy has a "cash value." At any time, the young man can cancel his policy and walk away with the balance in the account.

4. Sometimes an insurance company will offer a low interest loan to the policy owner, secured by the cash value in the insurance account. This is worth taking a minute to think about because even though the interest rate is low, remember that you're paying interest to get the use of your own money!

A real insurance company has operating expenses and needs profits. Also, the money being held is invested, and things can go from very well to very poorly. The insurance company is investing our young man's money, and all he can do is hope that it is investing wisely while keeping a close watch on its costs. And again, there are Life Tables that are much more specific and closely tailored to the statistics that pertain exactly to you—sex, race, medical history, and so on. Someone with a particularly dangerous career or someone with a serious congenital heart defect probably can't get life insurance, and a table that reflects these considerations will look considerably different from a table that includes "all comers."

10.10 BREAKING DOWN THE YEAR

This topic is a little math intensive and is not necessary for you to understand the rest of this chapter. The only sophisticated math, however, is handled using a spreadsheet function. If you're willing to "go with" the use of the spreadsheet function and a brief explanation of what's happening, this section will give you a little more insight into the workings of Life Tables.

For whatever reason, an insurance company decided it would like to be able to sell term policies for half years rather than years. In order to price these policies, it

Table 10.6　Excerpt from the Life Table for Men with Some Curve Fitting Data

Age	q	l	Number of dead	By fit	% Error
45	0.003735	94,154	5,846	5,854	−0.14
46	0.004071	93,803	6,197	6,196	0.02
47	0.004428	93,421	6,579	6,574	0.07
48	0.004806	93,007	6,993	6,989	0.06
49	0.005206	92,560	7,440	7,439	0.01
50	0.005648	92,078	7,922	7,926	−0.06
51	0.006121	91,558	8,442	8,449	−0.09
52	0.006594	90,998	9,002	9,008	−0.07
53	0.007045	90,398	9,602	9,604	−0.01
54	0.007488	89,761	10,239	10,235	0.04

needed Life Tables with half-year rather than full-year steps. For a crude approximation, it could have just split the q yearly values in half, for example, for a man,
$q_{50} = 0.005648$, so it could have used $0.005648/2 = 0.002824$ for the first half of his
fiftieth year and the same number for the second half. This is not a terrible approximation, but since the q values are climbing year by year in the 50-year age range,
the result would have been to overcharge for a first-half fiftieth year policy and to
undercharge for a second-half fiftieth year policy.

My first step in creating a more accurate set of q values is to create a notation.
The probability of a man dying sometime during his fiftieth year is q_{50}. I'll call the
new set of numbers r to distinguish these numbers from the q numbers. Remember
that there will be an r number for the first half of each year's age and a different r
number for the second half.

Table 10.6 starts with an excerpt from the men's Life Table. I'm only considering ages 45–54 and I'm only copying the age, q, and l columns. Remember that
column l is the number of men alive out of the original 100,000. In the next column,
I showed the number of men dead out of the original 100,000. This column was
generated by simply subtracting the numbers in column l from 100,000. Figure 10.6
shows a plot of the number of men dead column versus the man's age.

Spreadsheet programs offer the ability to "fit" a formula to a set of data. This
means that the spreadsheet program takes a formula that you choose from a list
of available formulas and adjusts the parameters in this formula to make a plot
of it look as though this formula had actually created the original data. Some art
is involved in the choice of formula. For the data shown in Figure 10.6, I chose
a "second-order polynomial" as my formula. This is the part you don't have to
worry about if you're uncomfortable with the math, but keep reading for the
results.

The spreadsheet program calculated the formula

$$\text{Number of dead} = 18.09\text{age}^2 - 1{,}304.1\text{age} + 27{,}906$$

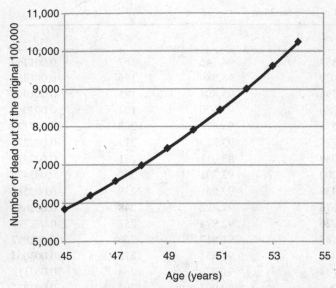

Figure 10.6   Curve fit used to generate a half-year Life Table section.

as its best fit to the data. In Table 10.6, I showed the formula predictions for the number of dead (labeled "by fit") and the percent error in the formula prediction versus the original data. In Figure 10.6, I have superimposed a graph of the formula on top of the graph of the original data. As you can see from the table, the percent error is very small. And as you can see from the figure, it's just about impossible to tell the difference between the two.

Table 10.7 shows how I built my new life table using my formula. The first column lists ages in half-year steps. The second column lists the number of dead generated by using my formula. The third column lists the number of alive (l), generated by subtracting each number of dead value from the original 100,000. The fourth column is the Life Table d column, the number of men dying each year. This is generated by subtracting the number of alive in one row by the number of alive in the previous row, for example, $d_{45.5} = l_{23} - l_{24}$.

Looking at the original Life Tables, d values are obtained by multiplying the number of people alive by the probability of their dying, $d_i = q_i l_i$ where i refers to any row (age) in the table. To get my new probability values, r_i, therefore, I just divide d_i by l_i. For example, $r_{45.5} = d_{45.5}/l_{45.5} = 176/93,980 = 0.001769$.[4]

The insurance company would no doubt do much more exotic curve fitting than I did and extend its results over the entire 0- to 100-year age span, but what I did above shows the calculations involved. I now have a Life Table that can accurately be used to prepare term policies in half-year increments.

---

[4] Again, I'm just showing rounded numbers here. The spreadsheet program is internally storing and using numbers to many more decimal places than shown. This creates some roundoff error in the results shown.

Table 10.7　Half Year Life Table for Men

Age	Number of dead	l	d	r
45.0	5,854	94,146	167	0.001769
45.5	6,020	93,980	176	0.001868
46.0	6,196	93,804	185	0.001968
46.5	6,380	93,620	194	0.002069
47.0	6,574	93,426	203	0.002170
47.5	6,777	93,223	212	0.002271
48.0	6,989	93,011	221	0.002374
48.5	7,209	92,791	230	0.002477
49.0	7,439	92,561	239	0.002581
49.5	7,678	92,322	248	0.002685
50.0	7,926	92,074	257	0.002791
50.5	8,183	91,817	266	0.002897
51.0	8,449	91,551	275	0.003004
51.5	8,724	91,276	284	0.003113
52.0	9,008	90,992	293	0.003222
52.5	9,301	90,699	302	0.003332
53.0	9,604	90,396	311	0.003443
53.5	9,915	90,085	320	0.003555
54.0	10,235	89,765	329	0.003669
54.5	10,564	89,436	338	0.003783

PROBLEMS

1. A-50-year old man decides that he wants a $1,000,000 life insurance for 2 years.
   (a) What will two separate 1-year term policies cost, both bought at the beginning of the first year? Use 4% annual percentage rate (APR) for calculating present values and ignore insurance company markups.
   (b) Repeat the above but treat this as a 2-year policy bought on the man's fiftieth birthday.
2. Suppose someone was to come down with an incurable terminal disease and was told by his or her doctor that "You have a two-third chance of lasting a year; you'll never last more than 2 years." Construct a Life Table for this person and calculate the cost of a $100,000 life insurance policy for this person.
3. Using the half year Life Table shown in Table 10.6, assuming an APR of 4% for the present value, what is the cost (idealized) of an 18-month, $50,000 term policy for a man starting halfway through his forty-eighth year?
4. Life Tables can be generated for longer time steps than 1 year as well as for shorter time steps than 1 year. Tables generated in 5-year steps are sometimes called abridged Life Tables. Starting with the women's Life Table, generate an abridged Women's Life Table.

5. You are taking a 5-year business loan for $250,000 that you will amortize yearly with 60 equal monthly payments at an APR of 10%. Your lender wants a fully paid-up life insurance policy to protect his or her interest. You will get a decreasing term policy that pays exactly the balance at the beginning of each year. You are a healthy, 35-year-old male.

(a) Create an amortization table for the loan and show the balances at the beginning of each year.

(b) Using the balances at the beginning of each year from problem 5a, create the 5-year decreasing term policy and get the price for this policy as an up-front payment. Use 5% as the APR for calculating present values. Assume that your insurance company prices this policy at 35% above the ideal calculation number.

(c) If you borrow the money for the policy along with your loan, what's your effective APR (effective annual percentage rate [EAPR])?

Chapter 11

Annuities

A single clear, concise, definition of annuity is impossible to create, because there are two principal financial vehicles called annuities and an untold number of variations and combinations within each of these two vehicle categories. A fixed annuity is considered in the United States to be a savings scheme with certain tax advantages and is regulated by the IRS. A variable annuity is considered in the United States to be an investment scheme, again with certain tax advantages, and is regulated by the U.S. Securities and Exchange Commission (SEC).

An annuity is a contract between you and an insurance company by which you give the insurance company money either in a lump sum or over time (the accumulation period) and then it starts sending you periodic payments. While the company has your money, it is investing this money and returning some of the profits (and possibly losses) to your account. The payments might continue for some fixed, specified, period of time or they might be for life.

A fixed annuity is an annuity that provides a guaranteed rate of return. The insurance company is taking the risk—it is betting that it can invest the money and earn a higher rate of return than the rate of return that it promised to give to you.

A variable annuity is an annuity in which the insurance company is investing the funds in your account for you and you are assuming both the opportunity for higher returns and the risk of investments. Variable annuities can get quite involved in that there might be a guaranteed part (i.e., a fixed annuity inside the variable annuity) and a host of subaccounts with different levels of risk and opportunity.

Annuity contracts can contain fees and/or periodic service charges. I'm not considering these costs here because they don't interact with understanding how the annuities work. Obviously, when shopping for an annuity, consider all of these costs and look for a (reputable) company with the lowest costs.

In the pages to follow, I'll be discussing some basic annuities. Because of the incredible number of variations available, I can't possibly discuss them all. My best advice is for you to read and/or make the salesperson explain things very, very, slowly until you're sure you understand exactly what you're buying. For example, there are rules concerning contract cancellation and return of remaining funds. Some

Understanding the Mathematics of Personal Finance: An Introduction to Financial Literacy, by Lawrence N. Dworsky
Copyright © 2009 John Wiley & Sons, Inc.

contracts just say no; some contracts allow for cancellation in the case of, say, medical emergencies. Fees are usually associated with these cancellations.

11.1 A BENCHMARK SAVINGS PLAN

In order to understand the advantages (and possible pitfalls) of a fixed annuity, it's useful to compare the annuity to a personal savings plan. The spreadsheet on my website Ch11FixedAnnuities.xls will help with these calculations. Go to the IAWPC tab (I'll explain this mysterious acronym soon). This example is shown in Table 11.1.

In this spreadsheet, I chose a starting month as month #1 so that the month number and the number of months gone by would be the same. The spreadsheet fixes the starting year at 1; I did this because we we're looking at years into the plan rather than actual dates.

I want to receive a payment of $2,500 each month, for 20 years, from this account. The annual percentage rate (APR) is 4.0%. My tax rate is 25%.

The calculated principal is $413,930 (top of column H). This is the amount that I have to deposit so that this account can fund my $2,500 monthly withdrawals for 20 years. Each month, the balance accrues interest. This interest will help to fund the plan. When I'm earning interest, I suddenly have a partner in this enterprise—the IRS. Savings bank interest is taxable. Each year, in month 4 (April), I have to pay income taxes.

With the tax rate entry set to 25%, the spreadsheet shows that in April of year 2, I must mail off $4,052 tax payment on this interest.[1] The tax bill column (column K) shows all the tax bills for the 20 years of the plan, and the PV column (column L) shows the present value of these bills reflected back to day 1.

For this example, the present value of the taxes is $37,322 (see the top of column L). You'll need this much available to pay for your taxes if you want this plan to be adequately funded to give you $2,500 a month for 20 years.

Having to take federal income tax into consideration forced an increase of $37,322 in principal. Now let's see if there's a way of saving some of this money.

11.2 IMMEDIATE ANNUITY WITH PERIOD CERTAIN

An immediate annuity is an annuity that starts making payments immediately. Period certain means that we're specifying how often the payments are made and how many payments there will be. This is the IAWPC acronym on the spreadsheet tab.

Once again, I am pointing out that anything I say about taxes might not be true when you're reading it. Tax laws change.

Assuming that things don't change drastically, the attraction of an annuity lies in how it is taxed. You pay taxes only on the money you pull out of the annuity (the

---

[1] There may have been estimated tax payments that were due along the way. I'm not going to worry about this here. I want to keep the calculations relatively straightforward.

Table 11.1a   Example of an IAWPC as Compared with a Savings Account

	Nr monthly payments:		240				
	Payment:		$2,500				
	Rate:		4%				
	Tax rate:		25%				

Savings:

Pmt Nr	Calculated principal: $413,929.83					PV of tax bills: $37,322	
	Mnth	Year	Balance ($)	Interest ($)	Tax	Tax bill	PV
1	1	1	411,429.83	0.00	0.00		
2	2	1	410,301.26	1,371.43	342.86		
3	3	1	409,168.93	1,367.67	341.92		
4	4	1	408,032.83	1,363.90	340.97		
5	5	1	406,892.94	1,360.11	340.03		
6	6	1	405,749.25	1,356.31	339.08		
7	7	1	404,601.74	1,352.50	338.12		
8	8	1	403,450.42	1,348.67	337.17		
9	9	1	402,295.25	1,344.83	336.21		
10	10	1	401,136.24	1,340.98	335.25		
11	11	1	399,973.36	1,337.12	334.28		
12	12	1	398,806.60	1,333.24	333.31		
13	1	2	397,635.96	1,329.36	332.34		
14	2	2	396,461.41	1,325.45	331.36		
15	3	2	395,282.95	1,321.54	330.38		
16	4	2	394,100.56	1,317.61	329.40	4,051.53	3,854.26
17	5	2	392,914.23	1,313.67	328.42		
18	6	2	391,723.94	1,309.71	327.43		
19	7	2	390,529.69	1,305.75	326.44		
237	9	20	7,450.28	33.06	8.26		
238	10	20	4,975.11	24.83	6.21		
239	11	20	2,491.69	16.58	4.15		
240	12	20	0.00	8.31	2.08	160.00	71.27

payments), not on the interest earned while the money is still in the account. This is called "tax-deferred growth."

In the case of the savings account, calculating the tax was easy. I just added up the interest earned each month for a year and then multiplied the result by the tax rate. In the case of this annuity, however, how do I calculate the amount of each

Table 11.1b   Example of an IAWPC as Compared with a Savings Account

Nr monthly payments:		240				
Payment:		$2,500				
Rate:		4%				
Tax rate:		25%				

Annuity:

Pmt Nr	Exclusion ratio:	0.690		PV:	$33,596	
	Mnth	Year	Taxable ($)	Tax ($)	Tax bill ($)	PV ($)
1	1	1	775.29	193.82		
2	2	1	775.29	193.82		
3	3	1	775.29	193.82		
4	4	1	775.29	193.82		
5	5	1	775.29	193.82		
6	6	1	775.29	193.82		
7	7	1	775.29	193.82		
8	8	1	775.29	193.82		
9	9	1	775.29	193.82		
10	10	1	775.29	193.82		
11	11	1	775.29	193.82		
12	12	1	775.29	193.82		
13	1	2	775.29	193.82		
14	2	2	775.29	193.82		
15	3	2	775.29	193.82		
16	4	2	775.29	193.82	2,519.70	2,397.01
17	5	2	775.29	193.82		
18	6	2	775.29	193.82		
19	7	2	775.29	193.82		
237	9	20	775.29	193.82		
238	10	20	775.29	193.82		
239	11	20	775.29	193.82		
240	12	20	775.29	193.82	2,325.88	1,036.08

monthly payment that's attributable to interest? There is no analytic way of deriving the answer to this—it requires a definition by the IRS.

The exclusion ratio is the fraction of the payment that is not taxed.[2] The numerator is the amount of money paid into the plan, which in this case is the principal.

[2] In this example, I am assuming that the principal is not being funded from some other tax-deferred source, for example, from an individual retirement account (IRA) or the like. In annuity jargon, I'm discussing a nonqualified annuity. Qualified annuities are annuities whose principal comes from funds that haven't been taxed (yet). They are taxed at higher rates than nonqualified annuities.

Since insurance companies sell these plans, you may see the principal referred to as the premium. The denominator is the sum of all the payments. For this example,

$$\text{Exclusion ratio} = \frac{\$413,929.83}{240(\$2,500)} = 0.690.$$

The fraction of each payment that's taxed is $1 - 0.690 = 0.31$. Given that all the payments are the same, all the tax bills are the same, as shown in columns N through O. In column Q, I showed the present value of these payments, and at the top of the column, I showed the sum of these present values ($33,596).

The present value of the tax bill for the annuity is definitely lower than the present value of the tax bill for the savings account. In addition, examining the tax payments shows that the savings account tax bills are "front loaded"—they are large in early years and small in later years. The tax payments of the annuity are the same year to year. In an economy where there is always some level of inflation, this means that a larger fraction of the annuity tax bills than of the savings tax bills is paid for with "cheaper" dollars. This can be a good thing for you, as was explained in Chapter 9.

11.3   DEFERRED ANNUITIES

The immediate in immediate annuity means that payments start immediately upon paying the premium. Often, the premium is accumulated by the purchaser over time. With a deferred annuity, the premium can be built up with a number of payments that grow with the tax on the interest deferred. This increases the growth of your payments as compared with putting the money into a savings bank. When payments begin, the exclusion ratio is lower than in the case of the immediate annuity, so that more of the payment gets taxed. The IRS is reclaiming some of the tax it deferred during the accumulation period.

Returning to the example above, I need a balance of $413,930 in my account on the day that I want the annuity to start making payments. I'll assume that I built this balance over the course of 10 years with regular monthly payments. First, I'll look at doing this in the savings account.

From the Deferred tab in the Ch11Fixedannuities.xls spreadsheet[3] or Table 11.2, I see that my monthly payment for the 10-year accumulation period is $2,811.07. I calculated the taxes on the savings and reflected them back to the beginning of the accumulation period. I then took the savings tax PV number from the IAWC (remember you have to enter this yourself), reflected it back from the end of the 10-year accumulation period to the beginning of the accumulation period, and added this to the total of the deferral period tax payment present values. The total in this example is $40,381. Summarizing, the present value of all the tax payments for 10 years of putting money into this savings account and then 20 years of pulling the money back

---

[3] I could have linked the Deferred tab sheet to the IAWPC tab sheet and had it pick up input information automatically. I decided, however, that this would be an endless cause of problems if/when users start modifying these sheets.

Table 11.2a Adding Deferred Principal Growth to the Table 11.1 Example

Payout:			Number of monthly payments:			240	
			Payout:			$2,500	
Accrual:			Number of monthly payments:			120	
			Final balance:			$413,930	
Rate:						4.00%	
Tax rate:						25%	
From IAWC calculation:			Savings tax PV:			$37,322	

Savings:

Pmt Nr	Mnth	Year	Balance ($)	Interest ($)	Tax ($)	Tax bill ($)	PV ($)
1	1	1	2,811.07	0.00	0.00		
2	2	1	5,631.52	9.37	2.34		
3	3	1	8,461.36	18.77	4.69		
4	4	1	11,300.64	28.20	7.05		
5	5	1	14,149.38	37.67	9.42		
6	6	1	17,007.62	47.16	11.79		
7	7	1	19,875.39	56.69	14.17		
8	8	1	22,752.71	66.25	16.56		
9	9	1	25,639.63	75.84	18.96		
10	10	1	28,536.16	85.47	21.37		
11	11	1	31,442.36	95.12	23.78		
12	12	1	34,358.24	104.81	26.20		
13	1	2	37,283.84	114.53	28.63		
14	2	2	40,219.19	124.28	31.07		
15	3	2	43,164.33	134.06	33.52		
16	4	2	46,119.28	143.88	35.97	184.97	175.97
17	5	2	49,084.09	153.73	38.43		
118	10	10	405,590.04	1,338.14	334.53		
119	11	10	409,753.08	1,351.97	337.99		
120	12	10	413,930.00	1,365.84	341.46	3,871.08	2,570.80

out again, at the beginning of this entire 30-year "project" is $40,381. Now I want to do the same thing for the annuity.

In the case of an annuity, I don't pay taxes on the interest accrued during the 10-year accumulation period. However, the exclusion ratio is no longer the balance at the beginning of the payout period divided by the sum of the payouts. The numerator is replaced by the sum of the contributions during the accumulation period (the basis). In this example, these numbers are

$$\text{Exclusion ratio} = \frac{120(\$2,811.07)}{240(\$2,500)} = 0.562.$$

Table 11.2b Adding Deferred Principal Growth to the Table 11.1 Example

Payout:	Number of monthly payments:	240
	Payout	$2,500
Accrual:	Number of monthly payments:	120
	Final balance:	$413,930
Rate:		4.00%
Tax rate:		25%
From IAWC calculation:	Savings tax PV:	$37,322

Annuity:

Pmt Nr	Mnth	Year	Taxable ($)	Tax ($)	Tax bill ($)	PV ($)
1	1	1	1,094.46	273.62		
2	2	1	1,094.46	273.62		
3	3	1	1,094.46	273.62		
4	4	1	1,094.46	273.62		
5	5	1	1,094.46	273.62		
6	6	1	1,094.46	273.62		
7	7	1	1,094.46	273.62		
8	8	1	1,094.46	273.62		
9	9	1	1,094.46	273.62		
10	10	1	1,094.46	273.62		
11	11	1	1,094.46	273.62		
12	12	1	1,094.46	273.62		
13	1	2	1,094.46	273.62		
14	2	2	1,094.46	273.62		
15	3	2	1,094.46	273.62		
16	4	2	1,094.46	273.62	3,557.01	3,383.81
17	5	2	1,094.46	273.62		
238	10	20	1,094.46	273.62		
239	11	20	1,094.46	273.62		
240	12	20	1,094.46	273.62	3,283.39	1,462.61

While the immediate annuity payments were taxed at a 31% rate, these deferred annuity payments are taxed at a 56% rate. When I reflected these payments back to the beginning of the accumulation period, however, I got a present value of $31,812. The value of deferring taxes is clear. The comments above about inflation also hold here.

11.4 LIFE ANNUITIES

A life annuity starts out the same way that an annuity with period certain starts out. You pay a premium—either immediately or accumulated over a period of time (deferred)—and then you start getting regular checks in the mail. There is no predetermined cutoff date for these checks. They keep coming as long as you're alive––hence the name "life" annuity.

The structure of life annuities has many variations. Perhaps, the most common is an annuity that funds a married couple. If I buy a life annuity for myself, the payments stop when I die. If I'm one of a couple, I (we) could buy an annuity that keeps funding until both of us are dead. The annuity could be structured so that the payment amounts are constant throughout, or drop by some amount when the first of us dies.

Let's look at an immediate life annuity for one person—fixed payments starting on day 1 and continuing monthly until that person's death. The first key question that must be answered before doing any calculations is just how much premium is needed (at a given interest rate) to fund the payments. Unlike the annuity with period certain, we don't know how many payments there will be.

I downloaded Figure 11.1 from the IRS website in February of 2009 (IRS[4]). According to the figure's title, it's an actuarial table for one life. Looking at the table, you can see ages for males and females and a column labeled "multiples." Multiples is the number of years that the IRS estimates this annuity will be paying. The name multiples relates to the number that multiplies the annual payout, which in turn is the number in the denominator of the exclusion ratio formula.

How did the IRS come up with this table? Look at Table 10.1, the 2004 Life Table for all men in the United States. Column E in the Life Table is the expected number of years that a man of a given age is most likely to (is "expected to") live. In other words, column E in the Life Table is the estimate of the number of years that a man of a given age buying an immediate annuity will be collecting payments from that annuity, that is, the IRS multiple.

Looking through these tables, you will see that the 2009 numbers are more optimistic (people are living longer) than the 2004 numbers, and that women continue to outlive men. In the examples to follow, I'm going to stick with my 2004 Life Table numbers just for consistency with the rest of the book. The multiple will be one of the input variables in my spreadsheet, so you can use whichever numbers you wish.

At this point, we can start appreciating why annuities are closely tied to insurance companies. Insurance companies are inherently in the business of generating probability (actuarial) tables for the policies they write, and they understand the importance of averaging over a large number of people and the variabilities involved. Pricing life annuity contracts involves the same kinds of calculations as does pricing a life insurance contract—you just think in terms of how long people will continue to live instead of when they are going to die.

Calculating the relationship between the amount of the payments, the number of payments, and the premium is no different from in the previous examples. The number of monthly payments is 12 times the multiplier. You need to know the person's age when he or she buys a life annuity—you don't for a period certain annuity.

If the purchaser of a life annuity dies significantly earlier than the Life Tables predicted he or she would die, the balance in the account is forfeited. This is, in a sense, unfair to the beneficiaries of the purchaser but is a necessary attribute of the

[4] IRS. General rule for pensions and annuities. Publication 939. http://www.irs.gov/publications/p939/ar02.html (accessed February 2009).

Actuarial tables

Table I (one life) applies to all ages. Tables II–IV apply to males ages 35 to 90 and females age 40 to 95.
Table I—Ordinary life annuities—one life—expected return multiples

| Age | | Multiples | Age | | Multiples | Age | | Multiples |
Male	Female		Male	Female		Male	Female	
6	11	65.0	41	46	33.0	76	81	9.1
7	12	64.1	42	47	32.1	77	82	8.7
8	13	63.2	43	48	31.2	78	83	8.3
9	14	62.3	44	49	30.4	79	84	7.8
10	15	61.4	45	50	29.6	80	85	7.5
11	16	60.4	46	51	28.7	81	86	7.1
12	17	59.5	47	52	27.9	82	87	6.7
13	18	58.6	48	53	27.1	83	88	6.3
14	19	57.7	49	54	26.3	84	89	6.0
15	20	56.7	50	55	25.5	85	90	5.7
16	21	55.8	51	56	24.7	86	91	5.4
17	22	54.9	52	57	24.0	87	92	5.1
18	23	53.9	53	58	23.2	88	93	4.8
19	24	53.0	54	59	22.4	89	94	4.5
20	25	52.1	55	60	21.7	90	95	4.2
21	26	51.1	56	61	21.0	91	96	4.0
22	27	50.2	57	62	20.3	92	97	3.7
23	28	49.3	58	63	19.6	93	98	3.5
24	29	48.3	59	64	18.9	94	99	3.3
25	30	47.4	60	65	18.2	95	100	3.1
26	31	46.6	61	66	17.5	96	101	2.9
27	32	45.6	62	67	16.9	97	102	2.7
28	33	44.6	63	68	16.2	98	103	2.5
29	34	43.7	64	69	15.6	99	104	2.3
30	35	42.8	65	70	15.0	100	105	2.1
31	36	41.9	66	71	14.4	101	106	1.9
32	37	41.0	67	72	13.8	102	107	1.7
33	38	40.0	68	73	13.2	103	108	1.5
34	39	39.1	69	74	12.6	104	109	1.3
35	40	38.2	70	75	12.1	105	110	1.2
						106	111	1.0
36	41	37.3	71	76	11.6	107	112	0.8
37	42	36.6	72	77	11.0	108	113	0.7
38	43	35.6	73	78	10.5	109	114	0.6
39	44	34.7	74	79	10.1	110	115	0.5
40	45	33.8	75	80	9.6	111	116	0

Figure 11.1   One person actuarial table from the IRS website (from IRS Publication 939, General Rule for Pensions and Annuities, http://www.irs.gov/publications/p939/ar02.html, accessed February 2009).

annuity if the insurance company is going to have the money to continue making payments to the purchasers who outlive their life expectancies. This issue can be mitigated by adding a period-certain feature to the life annuity.

With a period-certain life annuity, a certain number of payments, for example, 10 years, is guaranteed. If the purchaser dies earlier than his or her life expectancy, his or her beneficiaries continue to receive his or her payments until the end of the specified period. What's happening here is that the insurance company is selling both a life annuity and a decreasing term life insurance policy bundled together. The life insurance policy is written to fund the payments for the period certain of the annuity, if needed. Since life insurance policies cost money, a life annuity with a period-certain provision must cost more than a simple or "pure" life annuity. Alternatively, for the same cost, a life annuity with a period certain provision will make smaller payments than will a pure life annuity.

Table 11.3 Life Annuity Example

Age:	60
Multiplier:	24.2
Payment:	$150
Rate:	8.9%
Tax rate:	25%
Nominal number of monthly payments	290.4
Calculated principal:	$17,991.52
Exclusion ratio:	0.413
Expected age at death:	84.2
Tax fraction before expected death age:	0.587
Monthly taxes before expected death age:	$22.01
Monthly taxes after expected death age:	$37.50

In terms of taxation, the IRS isn't about to let the purchasers of a life annuity who live longer than the IRS expected them to live not pay taxes on their "extra" income. Starting out, the calculation of the exclusion ratio is the same as for period-certain annuities. Once enough payments have been made so that the full principal has been excluded, the exclusion ratio goes to 0. From then on, all payments are fully taxed as ordinary income. This switchover occurs at the expected death age.

The Life tab in the spreadsheet Ch11FixedAnnuities.xls is much simpler than the other tabs in this spreadsheet. That's because there's no need for a balance sheet or for present value calculations. The nominal number of payments is known once you enter the age and the multiplier (e in the Life Tables). This is the number of annuity payments that will be made if you die exactly on your statistical expected date of death. Enter the monthly payment amount and the interest rate that your balance will grow at and the principal (premium) for the annuity is known, as is the exclusion ratio. Knowing your tax rate (bracket), we then calculate the monthly tax. If you outlive your expected date, the full monthly payment is taxable as ordinary income—at your tax rate (Table 11.3).

11.5 PAYMENTS FOR COUPLES

In the case of a married couple, there are several options for structuring the payouts of a life annuity. The first option is what has already been discussed—a defined amount for the life of the annuity holder with or without a guaranteed minimum number of payments. The second option is a defined amount, with or without a guaranteed minimum number of payments, until both people have died. The third option is for a defined amount, with or without a guaranteed minimum number of payments, until one of the couple has died and then a second, usually lower, defined amount until the second of the couple has died.

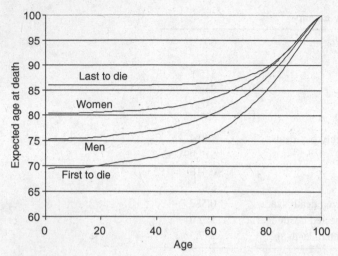

Figure 11.2 Expected life spans for men, women, and (same age) couples.

Figure 11.2 shows the expected life spans for men and women (taken directly from the Life Insurance chapter) and also two new curves: first to die and last to die. In this graph, the first to die and last to die curves were generated assuming that both the man and the woman are the same age. For same sex couples, there would be either a single men or a single women curve, surrounded by first to die and last to die curves.

As you can see from the figure, men statistically don't live as long as women. However, if we are just interested in when one of the couple will die—regardless of which one—then we can expect a younger demise than we would for either party individually. This result just arises from the probabilities: It's analogous to the probability of picking one green ball out of a bowl with one green ball and 1,000 red balls as compared with the probability of picking one green ball out of a bowl with two green balls and 1,000 red balls. When we raise the number of opportunities for our event of interest happening while keeping constant the number of opportunities for our event of interest not happening, the probability of our event of interest happening goes up. In this case, the probability of an early death increases because we are watching two people rather than just one person.

On the other hand, the expected age of the last to die goes up. The reasoning here is the same as the reasoning in the last paragraph. If I were to look at three people rather than two, I would expect to see an expected still younger first to die and a still older last to die. If I looked at, say, 10,000 60-year-olds, I would expect to see at least one of them dying this year and at least one of them lasting to 100.

Figure 11.2 is a snapshot of one out of all of the possible sets of curves; all couples are not of the same age. Table 11.4 is an abbreviated part of the table of tax multiples (or equivalently e values in the Life Tables) for man–woman couples of differing ages. What these tables tell us is the expected number of years from the

Table 11.4   First and Last to Die Multipliers for Couples

First	Men								
Women		50	55	60	65	70	75	80	85	90
	50	23.5	21.0	18.3	15.4	12.5	9.8	7.4	5.4	3.7
	55	21.7	19.7	17.3	14.8	12.1	9.6	7.3	5.3	3.7
	60	19.4	17.9	16.0	13.9	11.6	9.3	7.1	5.2	3.7
	65	16.8	15.8	14.4	12.8	10.8	8.8	6.8	5.1	3.6
	70	14.1	13.4	12.5	11.3	9.8	8.2	6.5	4.9	3.5
	75	11.3	10.9	10.3	9.5	8.5	7.3	5.9	4.6	3.3
	80	8.6	8.4	8.1	7.6	7.0	6.1	5.1	4.1	3.1
	85	6.3	6.2	6.0	5.8	5.4	4.9	4.2	3.5	2.7
	90	4.3	4.3	4.2	4.1	3.9	3.6	3.3	2.8	2.3
Last	Men								
Women		50	55	60	65	70	75	80	85	90
	50	36.9	35.3	34.1	33.4	32.8	32.5	32.3	32.2	32.2
	55	34.3	32.2	30.6	29.5	28.8	28.3	28.0	27.8	27.8
	60	32.3	29.7	27.6	26.1	25.0	24.3	23.9	23.6	23.5
	65	30.9	27.8	25.2	23.2	21.7	20.8	20.2	19.7	19.6
	70	29.9	26.4	23.4	20.9	19.0	17.7	16.8	16.2	15.9
	75	29.2	25.5	22.1	19.2	16.9	15.1	13.9	13.1	12.6
	80	28.8	24.9	21.3	18.1	15.3	13.2	11.6	10.5	9.9
	85	28.5	24.5	20.7	17.3	14.3	11.8	9.9	8.6	7.6
	90	28.4	24.3	20.5	16.9	13.8	11.0	8.8	7.1	6.0

start of a payout of an annuity when we can expect to see the first of the couple die and the expected number of years when we can see the second of the couple to die.

These tables are used in the following way. Suppose a 65-year-old man and his 60-year-old wife were to buy an immediate fixed annuity today. The annuity pays a certain amount monthly until the first of the couple dies, then pays, say, 75% of that amount until the second of the couple dies. We can think of what's needed as two separate annuities. First, we need an immediate life annuity starting today for the initial amount with an expected payout period of 13.9 years. Second, we need a deferred life annuity, starting 13.9 years from now, with an expected payout period of 26.1 − 13.9 = 12.2 years for 75% of the initial amount.

A defined benefit pension that some workers receive upon retirement looks a bit like the annuities above but has some differences. On retirement, the worker is usually given the options of either receiving a full amount for his or her lifetime or receiving a reduced amount until the last to die of the couple passes away. In the former case, the retiree is receiving a conventional immediate fixed annuity based on his or her expected date of death and the pension fund accrued for this worker. In the latter case, the reduced amount is based on the employer taking the same premium and figuring that it will be paid until the expected date of death of the sex of the second to die.

11.6 ONLINE CALCULATORS

There is no shortage of online annuity calculators. Due to the high number of combinations of attributes possible, you have to be very careful to understand just what you're calculating. Also, in most cases, these calculators are posted by companies that want to sell you something—in the case of annuities, a very expensive something with big commissions for the successful salesperson.

This link isn't to a calculator. It's to the SEC's primer on annuities. If you're considering an annuity purchase soon, this is very good (and trustworthy and current) reading: http://www.sec.gov/answers/annuity.htm.

The first link below is the most flexible calculator I've found. It's also pleasantly free of advertising materials:

1. http://www.1728.com/annuity.htm;
2. http://www.annuity.com/retirement.cfm;
3. http://www.annuity.com/tax-deferred.cfm;
4. http://www.immediateannuities.com/;
5. http://www.moneychimp.com/calculator/annuity_calculator.htm.

11.7 VARIABLE ANNUITIES

The variable annuity is a relative newcomer to the annuity stage. A variable annuity is an investment program tied together with the tax-deferred properties of an annuity. As when comparing investing to (savings bank) savings, the purchaser of a variable annuity is trading the opportunity of higher returns for a higher level of risk.

The subject of variable annuities gets incredibly complicated because of the multitude of options available to the purchaser. For example, there can be guaranteed minimum income amounts and the ability to switch to fixed payments or even to cancel out a contract (and of course pull your money out) once annuitization has begun. This last option actually translates to a tax-deferred long-term investment plan.

The purchaser of a variable annuity gets to choose investment portfolios, pretty much as if he or she were choosing mutual funds. A single annuity can usually participate in several different portfolios simultaneously, with the proportion of investment in each portfolio decided by the investor.

I can't offer anything in the way of a calculator for variable annuities because the *variable* part makes prior calculation impossible. It would be like asking for a spreadsheet calculator to predict the future of the stock market. If I could do that ... This is why the federal (U.S.) government regulates variable annuities through the SEC; the government considers these annuities as an investment plan. Since these annuities are sold by insurance companies, there will also be state regulation involved.

PROBLEMS

1. Today is your birthday. You are a 75-year-old single man. You have $500,000 in the bank and $35,000 in yearly taxable income. You decided to buy an immediate fixed annuity with a $350,000 premium, leaving the remaining $150,000 in the bank for emergencies. Assume that your savings bank pays 5% and your annuity pays 7.5% APR. What will your total after-tax income be? Round off numbers a bit for convenience. Also, assume that when you filled out your tax return, you found that your taxable income is $10,000 less than your total income and that you live in a state with no state income tax.

2. You are a 75-year-old woman married to a 70-year-old man. You want to buy an immediate fixed life annuity that will pay $2,500 a month (pretax) until the first of you dies, and then $2,000 a month until the second of you dies. The company you are dealing with promises 6% growth on all accounts. Estimate the premium (assume that today is your seventy-fifth and his seventieth birthday).

3. You have a paid-up whole life insurance policy for $250,000, listing your daughter as the sole beneficiary. You are an 82-year-old widower who could use some cash. Fortunately, your daughter is financially secure and would rather see you enjoy your old age than count on your insurance to fund her future. Could you structure a do-it-yourself annuity?

4. For most workers, the Social Security System provides a deferred life annuity. You and your employer pay into the system while you're working, and then when you retire, you receive a monthly check for the rest of your life. Social security payments actually receive a periodic cost of living (inflation) correction, but let's ignore both inflation and the correction in this problem. What I've read in several sources (e.g., http://bulletin.aarp.org/yourmoney/socialsecurity/articles/social_security__it.html) is that it's better to delay starting when a retiree should receive social security payments. According to these sources, for every dollar you receive monthly if you start at age 66, you only receive $0.75 if you start at age 62, but you receive $1.32 if you wait until age 70. Can you argue for or against the conclusion that you should wait until age 70? Assume you're in good health with no very unusual family history.

Chapter 12

Reverse Mortgages and Viatical Settlements

12.1 REVERSE MORTGAGES

A reverse mortgage is a variation on a home equity loan with some insurance and annuity concepts scrambled into the mix. Before going further, I should mention that government regulations limit the availability of reverse mortgages to people age 62 or over.[1]

Assume that you, or the youngest of you and your partner, is at least 62 years old. You own your home (or the two of you own it jointly). Your personal credit rating is unimportant except for a few specific items such as an outstanding tax bill.

A reverse mortgage will give you a lump sum loan today (or another option that I'll present later on). You don't make any payments on this loan. You continue living in your home until the second of you dies or you decide to sell the home. At that time, your beneficiaries can either pay off the loan (including accumulated interest) from their own funds or sell the home and pay off the loan with the proceeds. If the home has appreciated to a greater value than the amount due on the loan, the beneficiaries keep the difference. If the home has depreciated to a lesser value than the amount due on the loan, the lender absorbs the loss—at no time is more than the value of the home due to the lender.

For many people, this deal is wonderful. They get to spend the rest of their lives, or at least as much of the rest of their lives as they want to, in their own home. They do, of course, have to pay property taxes and insurance and maintain the home, but these aren't unreasonable requirements. They get some money that might be necessary or might just be used to make their later years more comfortable, but in any case, it's useful money. They have no risk of the deal collapsing and of their having either to come up with some money or to get out of their home. There is no cash risk to the beneficiaries.

What isn't there to like?

[1] Again, laws change. Check for current information.

Understanding the Mathematics of Personal Finance: An Introduction to Financial Literacy, by Lawrence N. Dworsky
Copyright © 2009 John Wiley & Sons, Inc.

- Scams and overly aggressive salespeople are associated with reverse mortgages.
- Start-up costs and administrative fees are high and, before regulation, some had been exorbitant.
- Elderly people are sometimes pressured to buy unnecessary insurance policies to accompany the reverse mortgage.
- The homeowners do not necessarily receive as much as they expect based on the appraisal of the home.

In putting together a reverse mortgage, the lender must first estimate when the homeowner (or the surviving spouse) will die. This is straightforward. In the last chapter, I presented couples' first to die and last to die tables to augment the individual Life Tables. Insurance companies are very good at making these estimates. The insurance company also needs to estimate the value of the home at the (estimated) time of that last death.

Most reverse annuity loans are adjustable rate mortgages (ARMs). The annual percentage rate (APR) can and almost definitely will vary. Depending on the program chosen, limits are set on how much and how quickly the APR can change.

When selling an annuity, an insurance company is looking to two sources of money to make payments on the annuity and run its business. First, it has the premium itself. Second, the company has the investment income it can earn on the balance of the premium as the company spends it down. When selling a reverse mortgage, the situation is different. The insurance company is handing money to a homeowner and is planning to recoup the money in the future when the home is sold. Conceptually, the insurance company is borrowing the money from somebody the day it gives the money to the homeowner and is planning to pay a lower interest rate than the rate it is charging so that it can run the business, make a profit, and settle the accounts when the home is ultimately sold.

What have I left out? I've left out some protection for the insurance company. What if home values aren't what everyone thought they would be when the time comes to sell the home? While excess home value goes to the beneficiaries, the insurance company doesn't collect the missing amount on insufficient home value. One of the costs of a reverse mortgage (which is wrapped into the loan) is an insurance policy for just this contingency. Various fees and administrative costs are also usually wrapped into the loan.

Reverse mortgages make more sense for very elderly people than they do for younger (but still eligible) elderly people. The shorter the time the insurance company (the lender) has to plan to wait before concluding the deal (i.e., the home is sold, the loan repaid, and the books are closed on the contract), the better its "crystal ball" predicts future home values, interest rates, and inflation, and consequently the more money it can make available for the loan.

A couple of possible variations on the theme: Remember that you don't have to die to get out of this. You can pay off the loan at any time. You can sell your house at any time but of course then you must pay off the loan. Next, you don't have to take the loan in a lump sum. You can structure a payment plan and/or you can use the loan amount as an available equity-based resource—you write a check on part or all of it when you need it. The advantage of either of these alternatives over taking the lump sum is that you don't pay interest on money you haven't yet borrowed. Since you'd undoubtedly be paying a higher interest rate on borrowed money than you could get by putting the lump sum into a savings bank, it's to your benefit not to pull out money until you need it or want to spend it.

There are several different programs and plans for getting a reverse mortgage. Each has its own specific details, its own costs, its own way of valuing your home, its own maximum loan amount, and so on. There is a tremendous amount of information and a calculator with a detailed calculation breakdown at http://www.reverse-mortgage.org/.

The examples shown in Table 12.1 were run on this calculator at the end of February 2009. I repeated these examples for homes in:

Shoreline, WA—a suburb of Seattle;

Scottsdale, AZ—a suburb of Phoenix;

Barrington, IL—a suburb of Chicago;

San Pedro, CA—a suburb of Los Angeles; and

Middlesex, MA—a suburb of Boston

and got the same answers each time. This is for the Department of Housing and Urban Development (HUD) Home Equity Conversion Mortgage (HECM) loan. The HECM is the most popular reverse mortgage program available today, but not the only available program.

The initial ARM rate on these loans was (approximately) 3.64%. This rate is based on the Treasury Notes, with the ARM adjustable monthly. A fixed rate loan

Table 12.1   Sample HECM Reverse Mortgage Estimates of Maximum Available Loan

Age	House value ($)		
	300	400	500
	Maximum available loan ($)		
65	163.1	220.7	279.3
70	176.9	239.0	302.1
75	191.1	257.8	325.5
80	206.3	279.9	350.5

All amounts are in $100,000 units.

Table 12.2   Sample HECM Reverse Mortgage Estimates of Maximum Available Loan,
Corrected for Bundled Costs

Age	House value ($)		
	300	400	500
		Maximum available loan ($)	
65	142	200	258
70	156	218	281
75	170	237	305
80	185	259	330
	Maximum available loan as a percentage of the house value (%)		
65	47	50	52
70	52	55	56
75	57	59	61
80	62	65	66

All amounts are in $100,000 units.

cost was approximately 6.2%. This number includes 0.5% mortgage insurance, so while the actual loan APR is 5.7%, you pay 6.2%.

The total of fees and service charges was about $21,000 for the loans. Table 12.2 is a repeat of Table 12.1 but with $21,000 deducted from each loan. Remember that these are approximate numbers; everything changes with time and not all the loans shown here have exactly the same bundled costs. Assuming you take a lump sum loan for the maximum available amount, you get the number in Table 12.2 while you repay from the number in Table 12.1.

Also in Table 12.2, I rewrote the numbers for what you actually get as a percentage of the value of the home. Table 12.2 is shown in a graph in Figure 12.1.

Figure 12.1 shows that the available loan as a percentage of the house value is greater for a more expensive home than for a less expensive home, but not by much. On the other hand, while only about 50% of the home's value is available to a 65-year-old borrower, about 65% of the home's value is available to an 80-year-old borrower. Also shown in Figure 12.1 (the dashed line with its axis on the right) is the e value from the Life Tables versus the age of the borrower at the time the loan is taken. The 65-year-old borrower expects to live about another 17 years while the 80-year-old borrower expects to live about another 8 years. The lender is more conservative with longer term loans, as was explained above.

A reverse mortgage is a loan issued by an insurance company. As long as the insurance company can charge a higher APR on this loan than it is paying for money "loaned" to it by other sources of income, such as life insurance premium income or by actual outside loans, this is a profitable business.

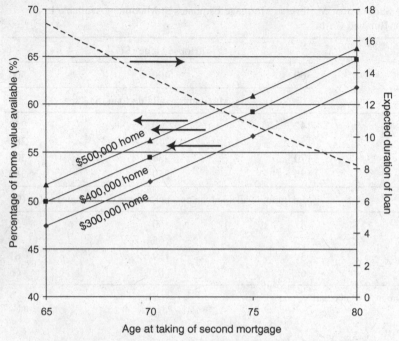

Figure 12.1   Sample HECM reverse mortgage estimates of maximum available loan, corrected for bundled costs and shown as a percentage of the house value.

I'm not offering my own custom calculator for reverse mortgages. The calculator referenced above is excellent, easy to use, and always current. The level of detail available is excellent. I can't do better.

12.2 VIATICAL SETTLEMENTS

A viatical settlement is a payment of a life insurance policy while the policyholder is still alive. These settlements are not available to the general public; they're restricted to people fitting specific requirements of having a terminal disease and/or short anticipated life span. The intent of the settlement is to help the policyholder handle the unexpected high costs of a very serious illness.

From a financial point of view, the insurance company has an expected date of death based upon medical prognoses and can easily come up with a present value of the policy at the time of writing the check to the policyholder. Unlike with a reverse mortgage, the insurance company has no ongoing relationship with the policyholder. When the check is received by the policyholder, the relationship between the policyholder and the insurance company ends because the policy has been paid off.

PROBLEMS

Reverse mortgages are complicated products with factors such as housing market values, interest rates, and so on, varying drastically by location and over time. Also, the laws governing them are still evolving. This problem is intended to show how a reverse mortgage works as a product for an insurance company and how the risks are really kept under control.

1. Suppose that there never had been such a thing as a reverse mortgage and that you, an insurance company business planner, just thought of the idea. Your task now is to explain your idea by means of a worked through hypothetical example, showing how the insurance company makes money and really isn't taking on much risk.

   Assume that your customer is a 77-year-old single man. He owns a home in a good neighborhood that appraises at $450,000. Your company can borrow money today at 5.00%, but there is some concern about rates changing, so you want to be able to write a contract with an adjustable interest rate. Your company's business model assumes a 2% spread between borrowing and lending to fund profits. Estimate your company's costs for preparing the package and servicing it as a $25,000 up-front fee. Looking at the Life Table for men (Table 10.1), a 77-year-old man is most likely going to live about another 10 years. Start by keeping interest rates constant at 5.00% so you'll have to charge your client 7%.

2. Using the calculator at the website http://www.reversemortgage.org/, run some estimates of your own. Vary the location, the value of the home, and the age of the borrower.

Chapter 13

Investing: Risk versus Reward

If you kept all your savings in cash in a shoe box under your bed, you would be risking loss due to theft, fire, and so on. On the other hand, while your savings would never add up to anything other than exactly what you put into the shoe box, from a financial point of view, your savings would be absolutely safe. No stock market variations, bank failures, or whatever could impact your savings. Inflation, however, would slowly eat into the actual value of these savings, even though the dollar amount didn't change.

Government-insured savings are, for all intents and purposes, perfectly safe. In addition, when your money is put into an insured saving product, you don't have to worry about a burglar making off with it in the night. Interest rates on these products are not very high, but they're usually a bit better than inflation. If total safety and security are your goals, then government-insured savings are the perfect choice. You won't get rich, but you'll stay ahead of inflation and you'll sleep well at night knowing that your money is secure.

At the other extreme, you could take all the money you want to put away for your old age and buy lottery tickets every week. In all likelihood, you're throwing your money down the sewer, but you never know; somebody will win the lottery and get very rich quickly. Chapter 14 discusses this approach to "investing" in more detail.

The holy grail of investors is the perfectly safe investment that returns a significantly higher average annual percentage rate (APR) than a savings account. Remember that a holy grail is something that people spend their lives looking for, but never find. They know it's out there somewhere, probably lying right next to the fountain of youth.

Before continuing on this topic, I should point out that I am writing this material early in March 2009. The stock markets are at their lowest levels in decades. Billions if not trillions of dollars have been lost, and nobody knows if the markets have reached bottom yet. Mainstays of American economic might such as General Motors are teetering on the brink of bankruptcy. Consequently, it's very hard for me to regurgitate the conventional wisdom about long-term growth of stock values being a historical truth.

Fortunately, I never intended to give stock market advice. For that matter, I've always wondered why people who know how to "pick good stocks" are spending their days being financial advisors for other people rather than picking these good stocks for their own accounts and quietly getting very rich. In an earlier book (Dworsky 2008), I discussed the concept of the "superior" fund manager and how it's fundamentally impossible to separate such a person from a crowd of randomly skilled fund managers. I'll go ahead and present some of the approaches to measuring and coping with investment risk, hoping that by the time you're reading these words, market stability has recovered. Keep in mind that when the stock market is rising, most portfolios will grow; when the market is falling, most portfolios will fall. If you're fortunate enough to be investing during a period of rising markets, be grateful for your earnings but be very cautious about attributing your success to your own skills.

There are so many different ways to invest money that I can't possibly present them all or give examples of more than a few of them in one short chapter. I'll treat a few concepts, mention some other ideas, and refer you to the public library sections on investing for the rest.

The general theme of investing is that potential reward comes with risk. There is some mathematical discussions of risk below, but the basic idea is that the faster you want your money to grow, the bigger the chance that something will go wrong.

13.1 STOCKS

There are many different investment products available today. Stocks and bonds relate to a specific company. You are buying a small part of or loaning money to the company of your choice. A share of stock represents ownership of a part of the company that issued the stock. As long as the company is out there doing business or at least still has some assets, your investment will have some value.

The risk in a stock price, given fairly calm market conditions, is measured by the stock's "volatility." In statistical terms, considering the stock price to be a random variable, the return on an investment has an expected value and a risk that is quantified by standard deviation. Expected value has been used several times already in this book. Standard deviation is a measure of how the stock's price (in this case) has varied about the expected value. I'll talk more about standard deviation below.

Figure 13.1 shows the performance of five hypothetical stocks over time for a period of 250 market days (about 1 year in actual time).[1] I've "normalized" the value of these stocks to a starting price of $1. That is, I've divided the price of each stock by its price on day 1. This lets me compare stock price change by percent changes as compared with their starting values. If I wanted to look at, for example, a stock selling today for $100 and a stock selling for $5, then it doesn't make sense to look at one share of each of these stocks. A $2 change in the $100 stock is noticeable but not exciting. A $2 change in the $5 stock is a major change in value. To compare these stock prices meaningfully, I have to look at equal dollar investments in each

---

[1] When I talk about a stock's price on a given day, I mean the closing price at the end of the day.

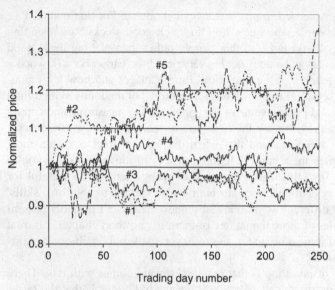

Figure 13.1   The price of five hypothetical stocks tracked for 1 year.

stock. I could compare the price of 50 shares of the $2 stock to the price of one share of the $100 stock, or I could look at what $1 would buy me of each stock, assuming that I could actually buy such pieces of shares of stocks. This latter choice is the mathematical equivalent of my normalization procedure.

At the end of the year, the average price of these five stocks is $1.105. If I had bought equal dollar amounts of all of these stocks on day 1 and waited a year, I would have earned slightly better than 10% APR on my investment. Had I put all of my money into stock #5, I would have earned about 37%. Had I put all of my money into stock #4, I would have lost about 4%. How do I know what to do?

I'll review standard deviation and also introduce the idea of correlation. These are statistical indicators that are very useful in guiding your investing decisions. They are not guarantees of anything.

Figure 13.2 shows histograms of stock #1 and stock #5. These are histograms of the 250 data points making up the stock's prices for the year. In the case of stock #1, all of the prices are bunched between $0.85 and $1.01, with the great majority bunched between $0.50 and $1.00. The average price for the year of this stock was $0.95. In other words, the stock price barely deviated from the average. The standard deviation of the stock price is a measure of how much the stock price deviated from the average. Without going through the math, the standard deviation for this stock was $0.030, approximately 3.2% of the average. The price from day to day probably reflected a bit of the company's business success level, a bit of overall market statistics, and a bit of random "noise." This is not a volatile stock.

Now look at the histogram for stock #5. The average price for the year of this stock was $1.12. The histogram shows stock prices varying from about $0.85 to about $1.35. Although most of the time the price was close to the average, there is

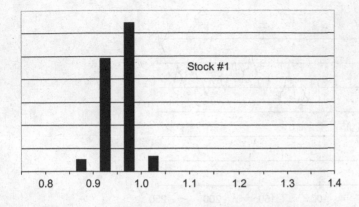

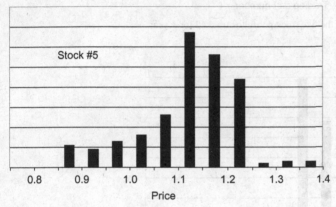

Price

Figure 13.2 Histograms of stocks #1 and #5 from Figure 13.1.

considerable spread, or width, to this histogram. The standard deviation is $0.10, about 8.9% of the average. This is a much more volatile stock than stock #1. Looking back at Figure 13.1, this stock was performing far worse at about 30 days into the year than any other stock in the group. Its price was less than $0.90 while the other four stocks were about either $1.00 or greater. There would have been a strong temptation to dump this "loser" and put the money back into the other stocks. At the end of the year, however, stock #5 was clearly the best performer of the group. The point here is that "more volatile" is another way of saying "less predictable."

13.2 PORTFOLIOS

The mix of stocks that you own make up what's called your stock portfolio. In this simple example, the portfolio consists only of the five stocks shown, with the initial investment being an equal amount of money put into each of these stocks. Figure 13.3 shows this portfolio's performance for the year, and Figure 13.4 shows the histogram of the portfolio's daily prices. The standard deviation here is 0.026, about

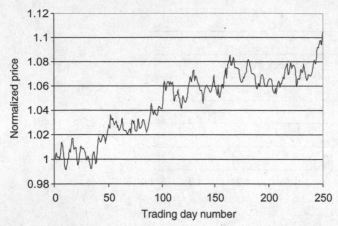

Figure 13.3   The price of the five stock portfolios tracked for 1 year.

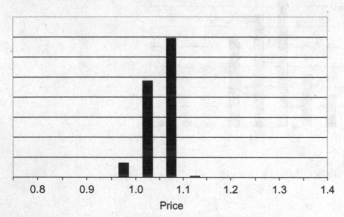

Figure 13.4   Histogram of the portfolio prices from Figure 13.2.

2.5% of the average. Even with the volatile stock #5 in the mix, the portfolio standard deviation is smaller than even the very nonvolatile stock #1 alone. The price of the portfolio at the end of the year is $1.05. The trade-off for this increased stability is the reduction of the highest attainable profit. While increasing stability is desirable, the possibility of giving up profit is unfortunate. What we have to do now is look at how portfolios are selected.

If you were to buy $1's worth of every stock in the stock market, your performance at the end of the year would be exactly that of the market itself. If you have faith in the future of the market and you want a very low-risk portfolio, this might be the way to go.

If you want to add some oversight that hopefully makes things better, you will probably pick a small number of stocks that you believe will outperform the market. To add some stability to this portfolio, you should diversify you holdings. Diversification is the management of correlation of stock performances.

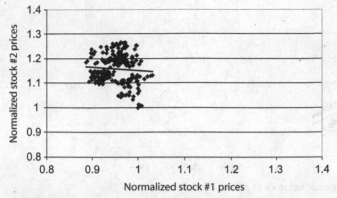

Figure 13.5 Scatter plot of stock #2 prices versus stock #1 prices.

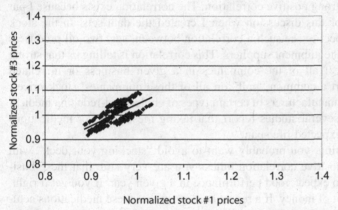

Figure 13.6 Scatter plot of stock #3 prices versus stock #1 prices.

In Figure 13.5, I plotted the (250) daily prices of stock #1 versus the daily prices of stock #2. The points don't seem to form a pattern, it's just a jumble of dots. The line represents the "best-fit" line to these points.[2] This line is nearly horizontal. If you recall the definition of the slope of a line, this line has a slope close to zero. The jumble of points and the horizontal line tell us that these two sets of data (stocks #1 and #2 prices) have almost nothing to do with one another; that is, they are uncorrelated. Buying stock in these companies is indeed diversifying your portfolio. You are randomizing your choices and are not letting the fortunes of one company be related to the fortunes of another company in your portfolio.

Figure 13.6 shows the same type of plot for stock #1 and stock #3. In this case, there is a recognizable pattern, and the best-fit line has a slope of almost +1. These

---

[2] Fitting of lines or curves to data is a fascinating topic that unfortunately is far outside the scope of this book. You can find references on the Web or in statistics texts under "fitting of curves to data" or "linear regression." Most spreadsheet programs offer a best-fit line generation option, usually as part of the graph generation commands.

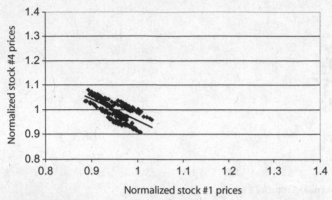

Figure 13.7   Scatter plot of stock #4 prices versus stock #1 prices.

two data sets show a strong positive correlation. The correlation exists because I put it there for the sake of this discussion when I created the data sets. In the stock market, you could expect to see such a correlation between, say, two oil producers or perhaps two medical equipment suppliers. This correlation is telling us that strong forces that might affect all of the companies in a given business or in related businesses would exert a common "pull" on all of these companies' stock prices. An example might be manufacturers of certain types of cholesterol-reducing medications in a year when several studies report that taking these types of medications really increases your expected life span.

In terms of investing, you probably want to avoid "stacking your deck" with companies having a positive correlation unless you are very sure that these businesses have a reason to expect good performance in a given year. If you get it right, you stand to make a lot of money. If a report comes out that these medications actually might do you harm, all of these drug companies will probably fare poorly for a while.

Figure 13.7 shows the correlation between data sets #1 and #4. Again, I created these data sets and the correlations for my examples—there is no deeper meaning here. In this figure, a pattern is once again obvious, and the best-fit line shows a slope of almost −1. These data sets are anticorrelated. These two companies might be a lambskin winter coat manufacturer and a light sweater manufacturer. A harsh winter would be good for the winter coat manufacturer, while a mild winter would keep people out of the winter coat manufacturer's stores and send them looking for light sweaters. In other words, there are factors that could drive one stock price up while simultaneously driving the other stock price down. Another example would be a manufacturer of a large SUV as compared with a manufacturer of a small hybrid car. Changing gasoline prices would significantly affect one of these businesses positively while affecting the other negatively. Buying stock in both of these companies might not make the most sense. If overall automobile sales are good, both companies could prosper—but a change in gasoline prices could cause one stock to go up while another stock goes down, and your portfolio just sees a profit cancelling a loss.

A simple way to diversify, rather than buying a few individual stocks, is to buy shares in a mutual fund. A mutual fund is a fund, managed by a professional manager, who buys selected stocks and then sells shares in the portfolio of these stocks. There are very many funds in the market. You can read about the "philosophy" of the fund manager for this particular fund. Many large fund companies offer several different funds, with choices in industries, risk level, social consciousness, "greenness," and so on, for the buyer to choose among. Since these funds are professionally managed, they're attractive for people who don't have the time, the expertise, or the interest in closely following the stock market. Also, because these funds are investing the pooled resources of many people, they can diversify to a level that most individual investors cannot.

13.3 CALCULATORS

The spreadsheet Ch13Stocks.xls will do correlation calculations and produces some numbers that I haven't introduced yet. To use this spreadsheet, you must enter some stock prices. You can do this by typing them in (I'll go through an example below) or if you're proficient enough with spreadsheet operations, download them from the Web right into the spreadsheet. An important point is that you must enter stock prices that are uniformly distributed in date. The closing price every Friday afternoon for a year, for example, is a good data set. Entering daily information for the month of January and then end-of-month information for the rest of the year won't yield any meaningful results. The spreadsheet has no way of knowing whether or not you're feeding it "good" data—it will try to do its thing with whatever you type in. Ch13Stocks.xls is deceptively simple-looking—there are a lot of calculations going on behind the scene.

The Correlation tab in Ch13Stocks.xls lets you look at two sets of stock prices and see how well they are or are not correlated. Table 13.1 shows a few examples of the use of this spreadsheet. These examples use a few data points—my purpose here is to show the use of the spreadsheet.

Examples I and II are the same for this spreadsheet. Order of data entry doesn't matter. You enter data to the left of the green line, as many data points as you wish, up to 1,000. If you don't enter the same number of points for both stocks, the message, "Please enter the same number of prices for both stocks" appears at the top of the spreadsheet.

The spreadsheet first reports the number of prices entered. Next, it shows the slope of the best-fit line, as in the example above. In this case, the slope is positive, indicating a positive correlation between the two data sets. Next, the spreadsheet reports the correlation coefficient. This coefficient will be positive if the slope is positive, and negative if the slope is negative.

As example III shows, the slope can be any number at all while the correlation coefficient can only vary between −1 and +1. In this instance, there is a fairly good correlation between the two data sets (+0.645), and the best-fit line predicts that stock #2, on the average, varied twice as much as stock #1. Goodness of fit is the

Table 13.1 Examples of Ch13Stocks.xls, Correlation Tab

Example I:
Stock prices:

Stock 1	Stock 2		
10.00	12.00	Number of prices entered:	3
11.00	10.00	Slope of best-fit line:	1.000
12.00	14.00	Correlation coefficient:	0.500
		Goodness of fit:	0.250
		Mean 1:	11.00
		Mean 2:	12.00
		Standard deviation 1:	1.00
		Standard deviation 2:	2.00
		(Standard deviation 1)/(Mean 1):	9.09%
		(Standard deviation 2)/(Mean 2):	16.67%

Example II:
Stock prices:

Stock 1	Stock 2		
10.00	10.00	Number of prices entered:	3
11.00	14.00	Slope of best-fit line:	1.000
12.00	12.00	Correlation coefficient:	0.500
		Goodness of fit:	0.250
		Mean 1:	11.00
		Mean 2:	12.00
		Standard deviation 1:	1.00
		Standard deviation 2:	2.00
		(Standard deviation 1)/(Mean 1):	9.09%
		(Standard deviation 2)/(Mean 2):	16.67%

Example III:
Stock prices:

Stock 1	Stock 2		
10.00	12.00	Number of prices entered:	3
12.00	16.00	Slope of best-fit line:	2.000
11.00	9.9	Correlation coefficient:	0.645
		Goodness of fit:	0.417
		Mean 1:	11.00
		Mean 2:	12.63
		Standard deviation 1:	1.00
		Standard deviation 2:	3.10
		(Standard deviation 1)/(Mean 1):	9.09%
		(Standard deviation 2)/(Mean 2):	24.53%

Example IV:
Stock prices:

Stock 1	Stock 2		
10.00	10.00	Number of prices entered:	7
11.00	11.00	Slope of best-fit line:	−0.393
12.00	6.00	Correlation coefficient:	−0.315
13.00	3.00	Goodness of fit:	0.100
14.00	8.00	Mean 1:	13.00
15.00	9.00	Mean 2:	7.71
16.00	7.00	Standard deviation 1:	2.16
		Standard deviation 2:	2.69
		(Standard deviation 1)/(Mean 1):	16.62%
		(Standard deviation 2)/(Mean 2):	34.88%

square of the correlation coefficient; hence, it's a number that varies between 0 and 1. If the correlation coefficient is 0, the goodness of fit is 0 and the two data sets appear to be totally independent of each other. If the correlation coefficient is −1 or +1, the goodness of fit is +1, indicating that there is excellent correlation between the two data sets. The sign of the correlation coefficient indicates negative or positive correlation. There was no need to show goodness of fit here, but it so often accompanies the fitting of a line to data that I thought I should include it and explain what it is.

Finally, example IV shows a set of data with a negative correlation coefficient (−0.315). The goodness of fit is only 0.100. These data are somewhat correlated; the points don't seem to "line up" at all (try graphing them).

All of the examples show the mean (average) of both stock prices, followed by their standard deviation. The standard deviations can be misleading because a $100 stock can have a higher standard deviation than a $2 stock while on a percent change basis not be varying as much. This is why I then show the standard deviations divided by (normalized to) the mean. These normalized standardized deviations are a measure of the comparative volatility of the two stocks.

13.4 DOLLAR COST AVERAGING

Figure 13.8 shows the daily price of three obviously fictitious stocks for about a year's trading. The stock labeled "going nowhere" starts at 100, then periodically cycles between 75 and 125, ending up at 100. The stocks labeled "climbing" and "falling" also start at 100 and cycle up and down, but end up at 120 and 80, respectively.

Table 13.2 shows the average price for these three fictional stocks and the average purchase cost per share if you bought the same dollar amount (this is *not* the same number of shares) every day for the year. On days when the stock price is higher than average, a dollar spent buys you comparatively less. On days when the

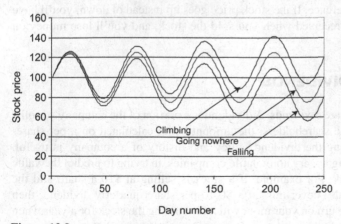

Figure 13.8   Dollar cost averaging example.

Table 13.2   Dollar Cost Averaging Examples

Stock	Average price ($)	Average purchase cost per share for $1.00 daily purchase ($)
Going nowhere	100	97
Climbing	110	107
Falling	90	86

stock price is lower than average, it buys you comparatively more. At the end of the year, the amount of stock you own is more than you would own if you had spent all of your money buying stock at the average price. This conclusion holds for stocks going nowhere, stocks that are climbing, and stocks that are falling. In the case of stocks that are falling, of course, you might wonder why you're buying them at all. In general, we invest in stocks because we believe that in the long run, values will grow and prices will climb. A healthy combination of purchase diversification and dollar cost averaging is perhaps an unexciting investment plan—but probably a very good one. In practice, dollar cost averaging usually means purchasing weekly or monthly or at your job's pay periods rather than the daily purchases shown in this example.

13.5 SHORT SALES

So far, this chapter presumed that the way to make money is to "buy low, sell high." This is true, but it isn't the only way to make money. Buying a stock in the hope that its price will rise is called "going long." Conversely, if you think a stock's price will fall, you can "go short" or "sell short." The way to do this is to arrange to borrow some stock that is selling at, say, $25 a share, and selling it. When the price falls to $20 a share you buy back the stock, return the shares and any costs for the loan, and keep the difference. If the stock price goes up instead of down, you'll have to pay more than you received when you sold the stock, and you'll lose money on the transaction.

13.6 STOCK DIVIDENDS

Corporations may declare dividends. A dividend is a portion of the company's profits that are returned to the shareholders, the amount being calculated on a per-share-owned basis. Examining the dividend policy and history of a company is useful, especially when looking at very nonvolatile companies, in trying to predict the value of the company's stock. For example, if a stock is selling at $10 a share and the company has historically never missed a $0.10 per share quarterly dividend, then you could earn a 4% return on your money just for holding the stock for a year when the stock price doesn't budge.

13.7 BONDS

A stock represents a part of a company owned by the stock purchaser. A bond, on the other hand, is a loan. A corporate bond is an "IOU" for money loaned to a corporation; a government bond is money loaned to the government. Various governmental bodies (federal, county, city) issue bonds. Because government bonds are considered to be very secure, they usually offer a lower interest rate than do corporate bonds.

The resale value of a bond depends on the interest rate the bond pays and the interest rates available elsewhere. If I have a bond that will repay $100 on its maturation date 1 year from today that is paying 5% annual interest, then I should be able to sell the bond today for the present value of the $100 repayment plus the present value of the year's interest.

Calculating the present value of course requires the assumption of an available interest rate. If that rate is very low, then the value of my bond today goes up. Conversely, if the available interest rate is very high, then the present value of moneys due to me a year from today goes down, and my bond value goes down with it.

Just as you can invest in mutual funds for stocks, you can invest in bond funds.

13.8 OPTIONS

Many types of "futures" products are available for the aggressive investor. I'm going to explain just one of them: options. A stock option gives the purchaser the right, but not the obligation, to buy (a "call") or to sell (a "put") a certain number of shares of a given stock for a preset price at any time up until the expiration date of the option.

That was a mouthful. Let me break it down, using the option to buy, or call, for an example. Widgetarama stock is selling at $28 (per share[3]) and I believe it will increase significantly in price over the next 90 days. I find that I can buy calls on Widgetarama stock for $4.00 with a strike price of $26 and an expiration date of 90 days from now. Exactly what am I buying? I am buying a contract that says that any time during the next 90 days, I can buy Widgetarama for $26, regardless of the actual selling price of Widgetarama on that date. The strike price is my guaranteed purchase price for the Widgetarama stock. The contract will specify how many shares I can buy. Actually, buying the stock at the strike price is called exercising my option. If I don't exercise my option before the expiration date, the option expires and I can toss it away. I don't get any money back.

Table 13.3 shows several examples of what might happen. I'm ignoring present value considerations so that we don't get buried in numbers. I do not intend to hold on to my shares of Widgetarama if I exercise the option; I'll sell the shares immediately and realize my profit.

---

[3] "Per share" is assumed when a stock or option price is quoted unless something else is specifically stated.

Table 13.3   Buying Calls Example

Stock price ($)	Profit ($)	Profit (%)	Annualized percent profit when exercised after (%)		
			30 days	60 days	90 days
29.00	−1.00	−25.00			
29.50	1.75	−12.50			
30.00	0.00	0.00	0	0	0
30.50	0.50	12.50	150	75	50
31.00	1.00	25.00	300	150	100
31.50	1.50	37.50	450	225	150
32.00	2.00	50.00	600	300	200

When I exercise my option, I'm paying $26 for a share of the stock. I've already paid $4 for the option. This means that the stock price has to be more than $30 for me to make a profit, or for the option to be "above water." If the stock price never climbs above $30 during the 90-day period, or I fail to exercise the option for whatever reason during this period, I lose money. This is a very important point. If I had bought shares of the Widgetarama stock and the stock price wandered around, up or down a few percent, the value of my investment would have wandered around this same few percent—I would have made or lost a few percent of my investment. If the option is never exercised, I've irretrievably lost my investment in the option. For stock prices between $26 and $30, I salvage part of my investment because the difference between my total cost and the price of the stock is still a loss, but it's less than $4.

This is a risky game to play. Why would I be doing it in the first place?

The answer to this question is in the table. I'm only showing results after 30, 60, and 90 days in the table. Suppose the stock price climbs to $31 sometime during the 90-day period. If I can buy a share of Widgetarama for $26 plus my $4 for the option and turn around and sell this share immediately, my profit is $30 − $26 − $4 = $1, which is 25% of my ($4) investment. I'm neglecting a small "operating cost" here: To buy the stock for $26, I had to first raise the money. Since I'll be selling the stock the same day, this is a very short-term loan. Even at credit card rates, a 1-day loan isn't that expensive. If I traded options regularly, I'd have a line of credit established and waiting for my needs. Also, there are commission costs any time a security is bought or sold.

If I had to wait 60 days for the stock price to reach $31, then my $4 has been tied up approximately one-sixth of a year, and my annualized profit is 6(25%) = 150%. At this time, my original investment and my profit are in my pocket and I can play the game again.

This is clearly an example of high-risk versus high-reward investing. If the stock price doesn't reach a high enough value during the lifetime of the option, I lose money, possibly everything. If the stock price climbs enough, I get a very high rate of return on my investment.

Table 13.4   Sample Actual Puts and Calls Listing

| Strike | April 2009 | | | | Oct 2009 | | | |
| | Call | | Put | | Call | | Put | |
	Ask ($)	Volume	Ask ($)	Volume	Ask ($)	Volume	Ask ($)	Volume
40	48.60	0	0.05	0				
45	43.60	0	0.05	0				
50	38.60	0	0.05	10	39.20	0	1.20	0
55	33.60	0	0.15	10	34.70	0	1.75	30
60	28.70	0	0.20	16	30.30	0	2.45	30
65	23.80	10	0.35	121	26.20	0	3.40	10
70	19.10	10	0.55	156	22.30	0	4.50	10
75	14.50	2	1.05	558	18.70	0	5.90	10
80	10.30	64	1.80	4,027	15.40	0	7.60	0
85	6.60	465	3.20	297	12.50	0	9.70	2
90	3.70	923	5.30	2,075	9.90	0	12.10	11
95	1.80	2,959	8.40	740	7.60	24	14.80	0
100	0.75	236	12.40	0	5.80	7	18.00	43
105	0.30	660	16.90	50	4.50	2	21.50	0
110	0.15	404	21.80	30	3.10	73	25.30	0
115	0.10	6	26.70	5	2.10	0	29.40	0
120	0.10	0	31.70	0	1.45	0	33.70	0
125	0.05	0	36.80	0	0.95	0	38.20	0
130	0.05	0	41.80	0	0.60	158	42.90	0

The Calls tab of the Ch13Stocks.xls spreadsheet lets you create tables such as Table 13.4. In the spreadsheet, Nr of Days is the number of days remaining before the option expires when you buy the option. In Table 13.4, I rounded 90 days to four times a year. The spreadsheet actually divides the number of days by 365, so the results look a bit different from the table results.

Option trading can get much more involved than simply buying calls. Let's look at some other option-trading opportunities.

You can sell calls. This means that you sell somebody a contract to buy a number of shares of a certain stock from you at a specified price and before a specified date—if he or she wants to. The call seller is the other party in the call-buying example above. When you sell a call, you receive a fixed amount of money. This money is yours to keep, whether or not the option is exercised. If the stock never gets above the strike price or the buyer never exercises the option for whatever reason, you walk away with this money. If the stock price goes up and the buyer exercises the option, you have to deliver the stock.

The potential risks and opportunities of buying and selling calls mirror each other. When you buy a call, you can never lose more than the price you paid for the call, but you can lose it all. If the stock price goes up enough, you get tremendous "leverage" and can make a lot of money. When you sell a call, you can never earn more than the price you received for the call. If the stock price goes up enough, your buyer can make a lot of money, which comes out of your pocket.

You can sell covered calls. Suppose you have some shares of Widgetarama that you bought a while ago. It's selling for $25 a share today; whether this is more or less than you paid doesn't matter right now. You sell somebody an option to buy these shares at a strike price of $27 for $5. If the option expires, that is, it is never exercised, you just earned $5. If the stock price goes up above $27, the option will get exercised and you must hand over your stock.

If the stock price never climbs above $27 before the option expires, you got your $5 for selling the option and you still own your stock. Even if the stock price falls below $5, you haven't lost money because of the $5 you were paid for the option. With a covered call, the stock price can go down by the amount you sold the call and you still make money (by selling the stock). If the stock price goes down even further, you will lose some money, but only the difference between the change in stock price and the amount you sold the call for. If the stock price goes up to the level where the option is exercised and you have to deliver the stock, you've still made money. What you've given up is the opportunity to make even more money because you're delivering the stock at a fixed price regardless of the market price of the stock.

You can buy and/or sell puts. A put is similar to a call except that with a put, you're buying or selling somebody the right to sell you a stock at a specified price.

By cleverly buying puts and calls at different strike prices, you can arrange things so that you make money if the stock either goes up or goes down by at least a certain amount.

Notice that there is no entry in the spreadsheet for the price of the stock when you buy an option. That's because this price doesn't affect any of the ensuing calculations. This does not mean, however, that this price is not an important number in your purchasing decision.

Table 13.4 is excerpted from the puts and calls listing for IBM stock on March 19, 2009 at midday. Notice how busy this table is—there are a lot of choices that can be made here. In Table 13.4, I'm only showing the April and October (2009) listings. The stock price at the time these listings were generated was 88.39.

First, look at the put and call asking prices for either the April or the October options. Note that the call price goes down as the strike price goes up, whereas the put price goes up as the strike price goes up. This is because the direction that the stock has to move for the option buyer to make money is opposite for puts and for calls.

Next, look at the volumes column. For both puts and calls in the April options, the curves are more or less "bell curves." Since these listings are from March 11, 2009, option buyers are developing pretty strong opinions about where they believe this stock will be selling at the end of April. In the October listings, however, the volume is fairly low and prices are all over the place. On March 11, the overall stock market is very volatile and people aren't sure about the future of the recession; they

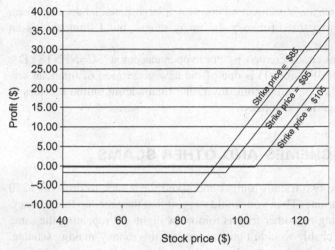

Figure 13.9   Profits and losses from the IBM option example.

are being conservative and there is no consensus about the stock market 7 months from now.

Figure 13.9 shows the trade-off between risk and potential reward, specifically what the future holds for the April $85, $95, and $105 IBM calls strike prices as a function of what the stock price is when the option is exercised (and the stock is sold). If you buy the $85 strike price option, you can lose as much as $6.60 (your option purchase price) if the stock price goes down enough. If you buy the $105 option, you can lose as much as $0.30, a far lower risk than losing $6.60. On the other hand, if the stock price goes up, you start making money sooner with the $85 option, and at all times (when you're making money), you make more money than you do with the $105 option.

13.9   ONLINE CALCULATORS AND LISTINGS

Stock, bond, mutual fund, and option listings are readily available from many sources. If you want to start trading, you can do it over the phone with a broker or online directly. You can get advice or you can be totally self-directed. There is a huge range of commissions (costs) for executing trades. Do your homework here.

The link http://finance.yahoo.com/ gives market updates and has a tab that offers quotes. Following this link will give you stock, bond, and option listings. There's a wealth of other resources available here, too.

The link http://www.aistockcharts.com/stock_correlation_tree_tool_help.htm offers a stock correlation tool. This link takes you to the introductory help page.

Vanguard (http://www.vanguard.com/us/FundsStocksOverview?gh_sec=n?WT. srch=1) is a mutual fund seller with many different offerings. I'm not suggesting that this is better or worse than many other companies, but I think that this is a very educational site.

Merrill Lynch (http://www.mldirect.ml.com/) is a large financial services firm. Again, I'm not promoting this firm over the many others, but I think this is an informative website.

E*Trade (https://us.etrade.com/e/t/welcome/openanaccount?SC=NPNT8YD& WT.mc_id=NPNT8YD&WT.srch=1) is one of the new generation of financial services firms that started as an online company rather than adding online services to a traditional financial services company.

13.10   PONZI SCHEMES AND OTHER SCAMS

Suppose that you called two friends tonight and asked them each to drop off $250 at your house in the morning. Then you would suggest that they recoup their money, and more, by each calling two other friends tomorrow night and repeating the same request to them. You probably wouldn't get too far with this money-making scheme.

A related scheme that has been used many times starts with an invitation from friends: "Come to a party Friday night and bring $100." Your name will be put in position 4 on a list. Then come the following Friday night and bring three friends, each with $100. They will be given the same instructions. Every week the names on the list are dropped one position and the people whose names are in position 1 split all the money coming in that night and their names get crossed off the list. You attend the party and watch all the people in position 1 receive $2,700 each. This isn't bad!

Week 2 you bring three friends with their money. Week 3 you just come to watch the fun. Then, finally, on week 4, you come to collect your $2,700. Will you really collect all of this money? Maybe.

The biggest problem is this: When you start, you are one of $1 + 3 + 9 + 27 = 40$ people in the room. The next week, there are $3 + 9 + 27 + 81 = 120$, followed by 360 and then 1,080. Can you picture the madness of showing up as part of a group of 1,080 people, 729 of them bringing money and 27 of them collecting money?

This type of scheme typically works early on, for the first few groups of people, and then falls apart as the number of participants get large. You could find yourself in the position of watching your next-door neighbor collect $2,700 while you never get to collect anything, or vice versa. In either case, the future of your relationship with this neighbor probably will not be good.

The above scheme is called a pyramid scheme. A more sophisticated scheme of this sort is called the Ponzi scheme, named after Charles Ponzi, an Italian immigrant to the United States in the early twentieth century. Ponzi schemes are devilishly clever in their ability to fool a lot of people for a long time. I'm going to work through a hypothetical example of a Ponzi scheme. It's a bit messy but worth your while to follow the money flow(s). Just in case you think this couldn't happen to you, note that in early 2009, a New York fund manager named Bernard Madoff was arrested for running a Ponzi scheme that fooled the most sophisticated investors and even fooled the Securities and Exchange Commission examiners for quite a while. Right now nobody is quite sure of the numbers, but it's estimated that he stole somewhere between $20 and $50 billion from his investors. (This is not a typo; he made off with *billions* of dollars.)

I'm making my example somewhat rigid in that everybody invests the same amount of money and each investment can start and end only on January 1. I'm doing this so that following the logic of what's happening doesn't get so entangled with the details of calculations that are certain to give you a headache. Keep in mind, however, that it's precisely these entanglements that help make a Ponzi scheme very difficult to spot. Even with this very constrained example, it's tricky to follow what's happening. As Mr. Madoff recently showed, a real working complex Ponzi scheme can be so intricate and entangled that even experienced professionals can't figure out what's going on.

Assume that I'm an experienced stock advisor and money fund manager with some level of credibility. This is necessary just to get peoples' attention.

I'm announcing a new investment fund. Each year, on January 1, I will accept 100 investors, each investing $100,000. I am promising a 25% annual return on invested money. An investor can pull out and withdraw all of his or her money on January 1, but only on January 1, of any year.

I know that I can get a 5% return on money quite safely.

I will assume that 10 people from each year's starting group of 100 will want to pull out each year, taking all of their money with them.

Something I'm not telling my clients is that, when they sign up and give me $100,000, I immediately take $25,000 and put it into my personal account. This account is hidden offshore somewhere—for reasons that will become clear soon.

Table 13.5 is very busy. That's because, since I'm setting up a scam, it's necessary to keep several sets of books. I need a book that shows my profits, a book that honestly tracks clients' money, and a book that I publish showing "how well" my fund is doing.

The first line in the table shows all my books at the beginning of year 1, the very start of the fund. My first group of 100 people has signed up. They each gave me $100,000, so I have $10,000,000 in my hands. Taking $25,000 from each of them gives me $2,500,000 in my own personal account and leaves $7,500,000 of clients' funds. I will spread this money over several bank accounts so nobody really knows how much money there is in the clients' fund accounts. Externally, each client believes that his or her account is now worth $100,000 and that the total fund is worth $10,000,000.

At the end of the first year, I do my own accounting. I still have only one group of 100 people. My personal account has grown at 5% to $2,625,000 and my internal (honest) record of clients' funds has grown, also at 5%, to $7,875,000. Externally, my clients each believe that their funds have grown to $125,000 each.

The next day is the beginning of year 2. Ten people pull out of group 1, leaving it with 90 people. My personal account doesn't change. My internal clients' funds account for group 1 drops by 10($125,000) = $1,250,000 to $6,625,000 because I returned the promised $125,000 to each of the 10 people who pulled out. At the same time, group 2 now joins the fund. I repeat everything I did with the money of group 1 at the beginning of the first year. That is, I take $2,500,000 for my personal

Table 13.5   Workings of a Hypothetical Ponzi Scheme

Time	Group	Number of people	My personal account ($)	Internal books ($)	External per person ($)	External total ($)
Beginning, year 1	1	100	2,500,000	7,500,000	100,000	10,000,000
End, year 1	1	100	2,625,000	7,875,000	125,000	
Beginning, year 2	1	90	2,625,000	6,625,000	125,000	11,250,000
	2	100	2,500,000	7,500,000	100,000	10,000,000
	Total	190	5,125,000	13,125,000		21,250,000
End, year 2	1	90	2,756,250	6,956,250	156,250	
	2	100	2,625,000	7,875,000	125,000	
	Total	190	5,381,250	13,831,250		
Beginning, year 3	1	80	2,756,250	5,393,750	156,250	12,500,000
	2	90	2,625,000	6,625,000	125,000	11,250,000
	3	100	2,500,000	7,500,000	100,000	10,000,000
	Total	270	7,881,250	19,518,750		33,750,000
End, year 3	1	80	2,894,063	5,663,438	195,313	
	2	90	2,756,250	6,956,250	156,250	
	3	100	2,625,000	7,875,000	125,000	
	Total	270	8,275,313	20,494,688		
Beginning, year 4	1	70	2,894,063	3,710,313	195,313	13,671,875
	2	80	2,756,250	5,393,750	156,250	12,500,000
	3	90	2,625,000	6,625,000	125,000	11,250,000
	4	100	2,500,000	7,500,000	100,000	10,000,000
	Total	340	10,775,313	23,229,063		47,421,875
End, year 4	1	70	3,038,766	3,895,828	244,131	
	2	80	2,894,063	5,663,438	195,313	
	3	90	2,756,250	6,956,250	156,250	
	4	100	2,625,000	7,875,000	125,000	
	Total	340	11,313,078	24,390,516		
Beginning, year 5	1	60	3,038,766	1,454,422	244,131	13,648,438
	2	70	2,894,063	3,710,313	195,313	13,671,875
	3	80	2,756,250	5,393,750	156,250	12,500,000
	4	90	2,625,000	6,625,000	125,000	11,250,000
	5	100	2,500,000	7,500,000	100,000	10,000,000
	Total	400	13,813,078	24,683,484		62,070,313
End, year 5	1	60	3,190,704	1,527,133	305,176	
	2	70	3,038,766	3,895,828	244,131	
	3	80	2,894,063	5,663,438	195,313	
	4	90	2,756,250	6,956,250	156,250	
	5	100	2,625,000	7,875,000	125,000	
	Total	400	13,504,782	25,917,659		

Table 13.5   Continued

Time	Group	Number of people	My personal account ($)	Internal books ($)	External per person ($)	External total ($)
Beginning, year 6	1	50	3,190,704	−1,524,615	305,176	15,258,789
	2	60	3,038,766	1,454,422	244,131	13,648,438
	3	70	2,894,063	3,710,313	195,313	13,671,875
	4	80	2,756,250	5,393,750	156,250	12,500,000
	5	90	2,625,000	6,625,000	125,000	11,250,000
	6	100	2,500,000	7,500,000	100,000	10,000,000
	Total	450	17,004,782	23,158,870		77,329,102
End, year 6	1	50	3,350,239	−1,600,846	381,470	
	2	60	3,190,704	1,527,133	305,176	
	3	70	3,038,766	3,895,828	244,131	
	4	80	2,894,063	5,663,438	195,313	
	5	90	2,756,250	6,956,250	156,250	
	6	100	2,625,000	7,875,000	125,000	
	Total	450	17,855,021	24,316,813		
Beginning, year 7	1	40	3,350,239	−5,415,543	381,470	15,258,789
	2	50	3,190,704	−1,524,615	305,176	15,258,789
	3	60	3,038,766	1,454,422	244,131	13,648,438
	4	70	2,894,063	3,710,313	195,313	13,671,875
	5	80	2,756,250	5,393,750	156,250	12,500,000
	6	90	2,625,000	6,625,000	125,000	11,250,000
	7	100	2,500,000	7,500,000	100,000	10,000,000
	Total	490	20,355,021	17,743,327		92,587,891
End, year 7	1	40	3,517,751	−5,686,320	476,837	
	2	50	3,350,239	−1,600,846	381,470	
	3	60	3,190,704	1,527,133	305,176	
	4	70	3,038,766	3,895,828	244,131	
	5	80	2,894,063	5,663,438	195,313	
	6	90	2,756,250	6,956,250	156,250	
	7	100	2,625,000	7,875,000	125,000	
	Total	490	21,372,772	18,630,493		
Beginning, year 8	1	30	3,517,751	−10,454,692	476,837	13,305,115
	2	40	3,350,239	−5,415,543	381,470	15,258,789
	3	50	3,190,704	−1,524,615	305,176	15,258,789
	4	60	3,038,766	1,454,422	244,131	13,648,438
	5	70	2,894,063	3,710,313	195,313	13,671,875
	6	80	2,756,250	5,393,750	156,250	12,500,000
	7	90	2,625,000	6,625,000	125,000	11,250,000
	8	100	2,500,000	7,500,000	100,000	10,000,000
	Total	520	23,872,772	7,288,635		106,893,005

account, which is now worth $5,125,000. I have $7,500,000 new clients' funds, bringing my clients' funds account up to $13,125,000. The group 2 people each have an account worth, so they believe, $100,000. I publish my funds report showing how much I paid out and what's left in the fund: $11,250,000 of first year investors' money plus $10,000,000 second year investors' money, totaling to $21,250,000.

The scam is evolving. Internally, I actually have only $13,125,000 in clients' funds, but I'm publishing a report showing $21,250,000 of clients' funds.

Following through the next few years, you can see the mess evolving. My personal account is growing healthily, the clients' funds account is growing slowly, and the published external funds total value is really looking fabulous! At the beginning of each year, 10 people from each group leave, and they each walk away with their original $100,000 compounded annually at 25%.

At the beginning of year 6, you can see the beginning of the end. The clients' fund value for group 1 (with only 50 people remaining in it) has gone negative. There isn't enough money to fund the necessary payouts. No one except me knows this. Because there is still plenty of money in the total clients' funds accounts, I just pay the missing piece out of other clients' money.

By the beginning of year 8, there isn't enough money in groups 1, 2, or 3 internal funds to pay the people pulling out. Since the total clients' fund account still has money, however, I can still pay them.

Let's take a minute to look at how things appear to the outside world at the beginning of year 8. The people who stuck with my fund for 8 years are each pulling out with $476,837. This is a 25% APR compounded for 8 years. The people who stuck with it for 7 years are pulling out with $381,470 and so on. These are very happy people. The published value of my fund going into year 8 is over $100,000,000.

Figure 13.10 shows the history (and the future) of my fund. To the outside world, it looks great. Internally, however, the end is in sight. There will not be

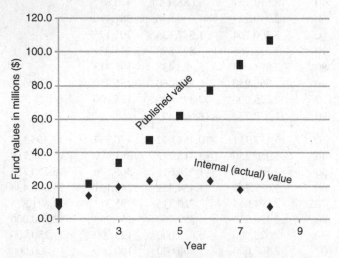

Figure 13.10   Published and internal Ponzi scheme fund values.

enough money to pay all the people who will be pulling out at the beginning of year 9.

At the beginning of year 8, my personal fund has almost $24 million in it. This isn't a bad retirement package. Since nobody expects to hear from me until the beginning of year 9, I have almost a full year to quietly arrange a new identity in remote but pleasant place in the world. A couple of million dollars goes a long way toward these goals. I might leave the $7.3 million in clients' funds, which is still scattered among a variety of banks, alone since I don't want to arouse any suspicions, or I might withdraw some or all of it. Then I disappear. Nobody notices that I'm gone until the beginning of year 9.

At the end of year 8, there are 540 investors, each believing their fund is worth something between $125,000 and about $600,000. These will not be happy people.

The scenario was idealized. I didn't account for my living expenses for the 7 years I was running the fund. I'd like to let a lot of people, whenever they show up, invest in the scheme. Not exactly 10 people will pull out from each group every year, and they certainly won't all want to pull out on exactly January 1. In a real fund, people will come and go at random times throughout the years. However, the workings of the scheme (or scam) are accurately portrayed. A lot of people made a lot of money (exactly what they were promised). They, unknowingly, helped the scam to succeed by building other investors' confidence that the fund was real.

A most unbelievable fact is that many people, including some professional fund managers, will watch their apparent fund value grow, year after year, at an incredible rate, without asking for a full annual report to see just what was happening. Bernie Madoff showed that they really will. Maybe many of them suspected what was going on but hoped that they had gotten in early enough to be one of the lucky winners. I'll never know.

I think the old adage that "something that seems too good to be true probably is" covers the ground very well here.

PROBLEMS

1. Suppose that you found three stocks that are ideal examples of prices cycling up and down similar to the dollar cost averaging example above. You are convinced that these three stocks average price at the end of the year will be the same as it was at the beginning of the year. The three variations, however, are significantly different. You wish to invest in each of these stocks, putting a fixed total amount of money in each month for a year. Would you be better off splitting your money equally among the three stocks or putting all your money into one of the three in terms of this idealized return?

2. The tab Sample Data in the spreadsheet Ch13Stocks.xls contains two columns of numbers that represent the prices of two stocks on 25 equally spaced (business) days. Copy these data into the stock price columns in the Correlation tab. Discuss the meaning of the various parameters presented.

   Replace the numbers in the stock #1 column with the numbers 1, 2, 3, ... 25. Can you find any meaning in the results?

3. The company you work for has given you some option shares (calls) as a perk with a strike price of $13 a share. Sketch a graph of your profits and losses and indicate where you would exercise the option.

4. Repeat the above exercise assuming you had purchased these option shares at $1 a share.

5. Going back to the situation of problem 1, you were given calls with a strike price of $13 a share. You now sell calls with a strike price of $18 a share for $1 a share. Repeat the above, assuming that your option buyer will exercise his or her calls when the stock price exceeds $18 a share and that both sets of calls expire on the same date.

6. You own both puts and calls for FrisbeesAreUs, Inc. The calls have a strike price of $90 and the puts have a strike price of $85. The stock itself is selling for $88. You paid $4 for the call and $5 for the put. Sketch a graph of your profits and losses as a function of the stock price and indicate where you'd exercise the put, the call, both, or neither.

REFERENCE

Dworsky, L. 2008. *Probably Not*. Hoboken, NJ: John Wiley & Sons.

Chapter 14

Gambling

You may have been told that any place you put your money that's not absolutely safe with insured protection is, to some extent, gambling. Accordingly, any investing, whatever the risk level, is gambling.

I'm going to be more restrictive. I'll only consider gambling to be betting your money on games of chance. Mixed systems, such as poker, combine skill and chance. You hope that your skill level will be higher than that of any of the other players so that in the long run, you will prevail.

Let me state my conclusion even at the start of the discussion. The only people who make money on games of chance are the casino owners. For games of chance, there are no systems; there are no lucky numbers; and there are no incantations—nothing can help you other than a little luck. And the little luck that you need is rare. I am always impressed by all the people who come back from a weekend in Las Vegas with more money in their pockets than they left with. I just can't understand how that poor town didn't go broke and shut down years ago.

14.1 PROBABILITY AND ODDS

Two equivalent measuring systems that are commonly used in gambling are the probability of winning (or losing) and the odds of winning (or losing).

If there are (W) ways of winning (say, the number of ways of pulling a green ball out of a box of 100 green and 300 red balls) and (L) ways of losing (the number of ways of not pulling a green ball, i.e., of pulling a red ball), then the probability of winning is the number of ways of winning divided by the total number of possible results. In the case of the box of 100 green and 300 red balls,

$$P = \frac{W}{W+L} = \frac{100}{100+300} = 0.25.$$

The odds of winning is the number of ways of winning divided by the number of ways of losing, usually expressed as a fraction with the word "to" designating division:

Understanding the Mathematics of Personal Finance: An Introduction to Financial Literacy, by Lawrence N. Dworsky
Copyright © 2009 John Wiley & Sons, Inc.

$$\text{Odds} = 1:3 = \frac{1}{3}.$$

In the red/green ball example, the following four statements are equivalent:

1. The probability of winning is 0.25 = 25%.

2. The probability of losing is 0.75 = 75%.

3. The odds of winning are 1:3, usually read as "1 to 3."

4. The odds of losing (or against winning) are 3:1, usually read as "3 to 1."

Converting between the two systems is easy:

$$\text{Odds} = P:(1-P)$$

or

$$P = \frac{\text{Odds}}{1+\text{Odds}}.$$

To use the formulas, the odds must first be written as a fraction, for example, $1:3 = 1/3$. When calculating odds from a probability, you'll get a fraction that will involve numbers less than 1. You have to scale the results up to integers before they look familiar. For example, a probability of 1/3 gives odds of $1/3:(1-1/3) = 1/3:2/3$. In this case, multiply the numerator and denominator by 3 to get the familiar result $1:2$.

In common jargon, when the probability of success is 0.5 = 1/2, the odds are $1:1$, called "even odds." When the probability of success is greater than 1/2, the first number in the odds is larger than the second number, for example, $3:2$. This is called "the odds are with you." In the opposite case, for example, $2:3$, the probability is less than 1/2, and "the odds are against you." When the probability of success is very small, the second number in the odds is much larger than the first number, for example, $2:15$, and the odds against you are "very long."

In dealing with odds, $1:3$ is the same as $2:6$ and as $3:9$, and so on. While expressing probabilities as odds may seem more intuitive to some people, it's a somewhat limited system. A probability of 0.12 (a little smaller than 0.125 = 1/8) is clear and accurate, but the equivalent odds of $12:88 = 3:22$ is an awkward representation. Something like 13:279 is really bad.

14.2 PROBABILITY AND EXPECTED RETURN

In the chapter on Life Insurance, I presented the idea of the probability of a random event (when you are going to die) and the calculation of an expected value (the average of when many thousands of people just like you are going to die). I'd like to extend these ideas to games of chance.

The expected value of my return on a game of chance is the sum of all the possible things that could happen multiplied by the probability of each of them happening. That sounds worse than it is. Look at the simple example of a coin flip game.

One of us flips a coin. If the coin lands heads up (heads), you give me a dollar; if it lands tails up (tails), I give you a dollar. In terms of the money in my pocket, giving me a dollar is +$1 and giving you a dollar is a dollar leaving my pocket, or −$1. When flipping a fair coin, the probability of heads is the same as the probability of tails, which is 0.5 Therefore,

$$E = 0.5(+\$1) + 0.5(-\$1) = 0.$$

The expected value of my return is 0. I should expect to neither win nor lose in the long run. In gambling terms, this is "even odds." I'll go through some coin flip game scenarios soon. Before that, I'd like to calculate a few more expected values.

Consider a roulette wheel. There are 38 slots. Each slot contains one of the following numbers:

$$0, 00, 1, 2, 3, \ldots, 36, 37, 38.$$

There are many different bets that can be placed, but I'll only consider one simple bet right now. You can bet that the ball will fall into an odd-numbered slot. This would be any of the slots 1, 3, 5, 7, ..., 33, 35, 37. Bet $1. If the ball falls into any of the odd-numbered slots, you get back $2, and you've won a dollar. If the ball falls into any of the even-numbered slots or the 0 or 00 slots, you've lost your dollar.

Since there are 38 slots, 18 of which are odd-numbered slots, your probability of winning is 18/38 and your probability of losing is 20/38. The expected value of your winnings on a $1 bet is therefore

$$E = \frac{18}{38}(+\$1) + \frac{20}{38}(-\$1) = -\frac{2}{38}(+\$1) = -\$0.053.$$

The expected return is negative. In gambling terms, the odds are against you.

Before examining just what this means, I'd like to present one more expected value calculation. Suppose that a local charity group has received a donation of a fancy TV that's worth $2,100. It plans to raffle this TV off. It will sell only 2,000 tickets at $1 per ticket. What's the expected value of a ticket purchase?

If 2,000 tickets are being sold and you buy one ticket, your probability of winning is 1/2,000 = 0.0005. Your probability of losing is 1 − 0.0005 = 0.9995. The expected value of your return on this gamble is

$$E = 0.0005(+\$2,100) + 0.9995(-\$1) = \$1.05 - \$0.9995 = +\$0.0505.$$

This expected value is positive. The odds are with you on this one.

Summarizing the results of the above three examples, I have a $1 bet on:

1. a coin flip game with an expected return of 0,
2. a bet on getting an odd number in a spin of a roulette wheel with an expected return of −$0.053, and
3. a raffle ticket for a $2,100 TV set with an expected return of +$0.0505.

These are three gambling opportunities with even odds, odds against me, and odds with me, respectively.

In order to study these three gambling games, I need to look not only at the probability of winning and the expected value of winning, but also at a very important third factor—how much money I have in my pocket when I start placing my bet(s).

First, let's look at the coin flip game. The expectation value is 0. You're as likely to win as you are to lose—or maybe not. The probability of getting heads (or tails) on a given coin flip is 1/2. Winning money with this game is a different story. If the game is to place a bet, flip a coin some number of times, and if you get more heads than tails then you win the bet, otherwise you lose the bet, then, you're as likely to win as you are to lose. If the game is as described above, you win $1 for heads, you lose $1 for tails; we have to look into some details.

Figure 14.1 is a "tree" of possible scenarios when you start out with $1 in your pocket. Each circle represents a coin flip. The number inside the circle is the amount in your pocket when you bet $1 and flip the coin. You have a 50% probability (even odds) of winning (*W*) or losing (*L*). If you lose (the arrow to the left), you have no money left; you can't bet again and you must quit. If you win flip #1, you have $2 and you can continue to bet.

The figure shows that the number of possible scenarios increases with each level of flip. At the first flip, there's a 50% chance of being wiped out and a 50% chance of increasing your holding to $2. At the second flip, you can't get wiped out; your

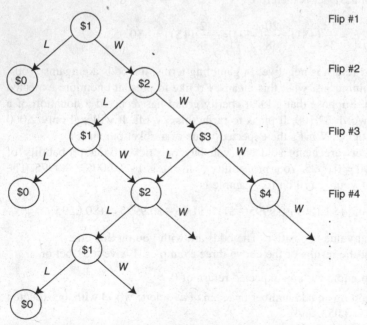

Figure 14.1 Probability tree for a coin flip game starting with $1.

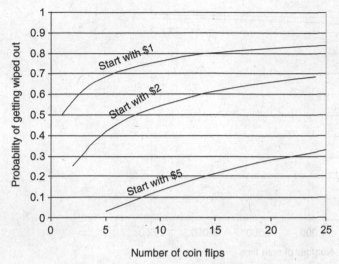

Figure 14.2   Probability of getting wiped out versus the number of coin flips—zoom in.

bankroll goes either down to $1 or up to $3. At the third flip, things start to get busy. Notice that at every flip level, there is a path to getting wiped out.

If you were to start with $2 in your pocket, the same chart works—just put yourself immediately at the flip #2 position and continue onward.

Figure 14.2 shows the total probability of getting wiped out versus the number of times you flip the coin (and bet $1). The more money you start out with, the lower your probability of getting wiped out. All of the curves shown, however, are increasing with the number of coin flips.

Figure 14.3 is the same as Figure 14.2 except that the probability of getting wiped out for up to 2,500 coin flips is shown. Starting out with more money helps, but no matter how much you start out with, sooner or later you will get wiped out. These curves assume that every time you win a dollar, you put it in your wallet and use it to continue betting. If you have to pull some money out for luxuries (dinner), then the probability of your getting wiped out sooner increases.

Another way of looking at this is to calculate the probability of your doubling your money before you get wiped out. Starting with $1, this is easy—you flip the coin and you've either doubled your money or you're wiped out. Interestingly, this is always the case. It doesn't matter how much money you start with. You place your dollar bet, flip the coin, and continue. The probability of your getting wiped out before you double your money stays at 50%. In other words, unless you like dragging things out, you might as well bet all of your money on one flip of the coin.

The probability of tripling your money before getting wiped out is 33.3% (1/3), the probability of quadrupling it is 25%, and so on.

Now let's go back to the roulette wheel. This time it's easy to reach a conclusion. The probability of winning is less than 50%. The expected value of return on a bet is negative. This can't possibly be a way to earn a living.

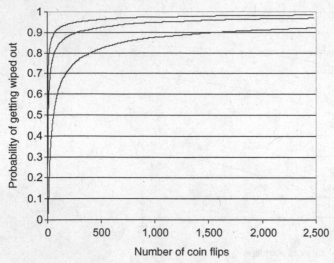

Figure 14.3 Probability of getting wiped out versus the number of coin flips—zoom out.

If there were a roulette wheel with 2,000 slots and you bet on any one of the slots, eventually you'd win. If you do this 1,000 times, then the probability of your winning at least once is about 39%. If you do it 2,000 times, then your probability jumps to 63%; for 5,000 times, it jumps to about 92%. The expected values here are a bit messy to calculate because if you bet 2,000 times, there's a probability of your winning once, a probability of your winning twice, and so on. Remember, however, that the expected value is always negative. No matter how much you start out with, you should expect to eventually be wiped out.

Finally, the TV raffle. Here, the probability of winning is very small (1/2,000), but the expected value of return is positive. If the expected value of return is positive, shouldn't you be able to use this as an income stream in the long run?

In the raffle example, the amount of money you start with in your pocket doesn't just warp the shape of a curve while still leading to the same conclusion; it can actually force different conclusions. First, suppose you have only $1. You buy a raffle ticket. Since the probability of winning is only 1/2,000, you almost invariably lose.

The raffle is different from the imaginary roulette wheel. Each ticket has a 0.005 probability of winning, and each ticket will pay $2,100 for a $1 bet. The roulette wheel never has to pay off. The raffle always has to pay off. With the raffle, if you could rush in early and buy all 2,000 tickets, then you walk away with the prize—there's no probability to calculate.

The bottom line here seems to be that in a roulette wheel gambling world where there is no required payoff, not only would you need a positive expected value of return, but you would also need enough starting money in your pocket to ride out the variations in order to stick around for a while.

In the real raffle world, there are times when the expected value is positive. The example of a donated raffle prize that's worth more than the sum of the cost of all the raffle tickets is such an example. If you buy a small fraction of the tickets sold, you'll probably lose your money. If you can buy all, or almost all, the tickets, you'll probably get a prize worth more than you paid.

Unfortunately, in most raffles and lotteries, the expected value of return is negative. It's not impossible to win (someone always wins), but it's simply a very poor way to spend your money for anything other than charity donations and possibly a little entertainment.

In real roulette wheel and slot machine gambling games, the expected value of return is negative. Occasionally, someone wins. Gambling casinos with a few hundred busy slot machines in one room often have each machine ring some loud bells when a player wins. The resulting cacophony makes it sound like winning is to be expected. Think about it. If the casino didn't see a positive expected value of return—equivalent to a negative expected value of return for the players—would the casino still be there week after week?

Also, assuming everything is honest, there are no systems. There is no such thing as a slot machine that's "due" to pay off because slot machines, like flipped coins, have no memory. There are no good lottery numbers. There are no bad lottery numbers. There are no lucky numbers. For a five-digit number, 11111 is as good as 12345 is as good as 42296 is as good as the number that won last week. People used to dealing in probabilities and random events often refer to lotteries as a "tax on stupidity."

14.3   PARI-MUTUEL BETTING

The pari-mutuel machine system used at race tracks offers a different approach to gambling. A roulette wheel owner is never sure what his or her daily operating cost will be; he or she only knows statistical generalities. On some nights, almost no one could win a lot of money; on some nights, there could be several lucky people. Also, the probabilities are fixed going into the game. The probability of the roulette wheel ball landing in any one slot is just 1/(number of slots) and the payoffs on any type of bet are published. The pari-mutuel machine system, on the other hand, adjusts the payoffs based on the bets, and the track owner can take a fixed amount or a percentage of the bets from each race. The easiest way to explain this is to work through an example.

Look at Table 14.1. The numbers are a bit contrived to make the example easy to follow, but the same calculations will work for any numbers. Six horses are racing, and people have placed bets on these six horses. Horse number 1 has almost one-third the total dollar amount of bets placed on it and is clearly "the favorite to win." Horse number 6 has very little money bet on it and is a "long shot" to win. The total amount of bets placed is $10,000.

This race track management has decided to distribute 90% of the bets to the bettors, keeping 10% for costs and profits. This leaves $9,000 to distribute. The numbers in the column labeled multiplier are calculated by dividing the amount to distribute by the amount bet on each horse: $9,000/$2,200 = 4.1; $9,000/$1,450 = 6.2; and so on. If horse number 1 wins, each dollar bet returns $4.10. If the favorite wins,

Table 14.1   Pari-Mutuel Betting Example

Horse number	Bets ($)	Multiplier
1	2,200	4.1
2	1,450	6.2
3	1,100	8.2
4	440	20.5
5	1,140	7.8
6	3,660	2.5
Total bets	10,000	
90% Payout	9,000	

Table 14.2   Pari-Mutuel Betting Example Including Winning and Placing

Winning horse #	Placing horse #	Equation relating bets to payoffs
1	2	$P_1N_1 + Q_1M_1 + Q_2M_2 = T$
1	3	$P_1N_1 + Q_1M_1 + Q_3M_3 = T$
2	3	$P_2N_2 + Q_2M_2 + Q_3M_3 = T$
2	1	$P_2N_2 + Q_2M_2 + Q_1M_1 = T$
3	1	$P_3N_3 + Q_3M_3 + Q_1M_1 = T$
3	2	$P_3N_3 + Q_3M_3 + Q_2M_2 = T$

each dollar bet on it returns \$2.50. If the long shot wins, each dollar bet on it returns \$20.50 and so on.

The next set of calculations is not necessary for understanding the basic idea of pari-mutuel betting. It's intended for those with some linear algebra background who are interested in how more involved pari-mutuel calculations are made. If you don't like the algebra, skip over it and just take a look at the examples.

At a race track, where you can bet on a specific horse winning (coming in first), placing (coming in first or second), or showing (coming in first, second, or third), the calculation of payback is, in principle, the same as above but gets a bit more complicated. I'll set up the case for just winning or placing and limit the race to only three horses.

In a three-horse race, there are six possible sets of results. Table 14.2 shows these six situations, with the horses labeled #1, #2, and #3. The table just shows the horse who won and the horse who placed—clearly the remaining horse finished last.

Let

N_1 = the number of dollars bet on horse #1 to win,

M_1 = the number of dollars bet on horse #1 to place,

N_2 = the number of dollars bet on horse #2 to win, and so on,

P_1 = the payoff multiplier for a winning bet on horse #1,

Q_1 = the payoff multiplier for a placing bet on horse #1, and so on, where payoff multiplier is the number of dollars paid for each dollar bet. Let T be the total amount bet, which is calculated by adding $N_1 + N_2 + N_3 + M_1 + M_2 + M_3$.

The amount of money paid to people who bet that horse #1 would win if it does win is $P_1 N_1$. The amount of money paid to people who bet that horse #1 would place if it wins is $Q_1 M_1$. The amount of money paid to people who bet that horse #2 would place if it wins is $Q_2 M_2$. The total amount of money paid out for the first line of Table 14.2 (horse #1 wins and #2 places) is therefore

$$P_1 N_1 + Q_1 M_1 + Q_2 M_2 = T.$$

(I'm ignoring the track owner's cut of the money.) Six such equations, one for each of the six possible results of the race, are shown in the table.

In algebraic terms, using matrix notation, this is a set of six coupled linear equations in the variables P_i and Q_i,

$$\begin{bmatrix} N_1 & 0 & 0 & M_1 & M_2 & 0 \\ N_1 & 0 & 0 & M_1 & 0 & M_3 \\ 0 & N_2 & 0 & 0 & M_2 & M_3 \\ 0 & N_2 & 0 & M_1 & M_2 & 0 \\ 0 & 0 & N_3 & M_1 & 0 & M_3 \\ 0 & 0 & N_3 & 0 & M_2 & M_3 \end{bmatrix} \begin{bmatrix} P_1 \\ P_2 \\ P_3 \\ Q_1 \\ Q_2 \\ Q_3 \end{bmatrix} = \begin{bmatrix} T \\ T \\ T \\ T \\ T \\ T \end{bmatrix},$$

that is easily solved by any number of standard techniques for the six unknowns, P_i, and Q_i. The payoff amounts may be scaled by any fraction to account for the track owner's cut, that is, if the track owner takes 10% then scale the results by a factor of 0.9.

Table 14.3 shows several examples of winning and placing of bets. I've used very small dollar amounts for these examples. Remember however that everything scales; the resulting multipliers for everybody betting a total of $10 are the same as the multipliers for everybody betting a total of $1,000,000.

In the first two examples, all the winning bets are the same and all the placing bets are the same. The result is that someone betting to win gets three times his or her money back, while someone betting to place only gets 1.5 times his or her money back. This makes sense. When betting to win in this situation, exactly one-third of the bettors will collect; when betting to place, two-third of the bettors will collect.

The third example is just a check on the calculation. In this example, no one bets to place, three-sixth of the bets are on #1 to win, two-sixth on #2 to win, and one-sixth on #3 to win. The resulting multipliers duplicate the simple situation already described.

In the fourth (last) example, I kept the winning bets the same as in the third example but changed the placing bets. Looking at the results, you'll see that even though the winning bets stayed the same, the winning multipliers changed. This is because the placing bets shifted the overall odds of the race, and the winning multipliers had to adjust accordingly. There is no obvious relationship between the winning multipliers and the placing multipliers, but the overall logic of betting on a horse that fewer people think will do well give you higher odds (a higher multiplier) and betting to win is still a riskier bet than betting to place, so the winning

Table 14.3   Some Examples of Pari-Mutuel Betting with Winning and Placing Bets

Horse #	Total dollar bet to win ($)	Total dollar bet to place ($)	Winning multiplier	Placing multiplier
1	1	1	3.00	1.50
2	1	1	3.00	1.50
3	1	1	3.00	1.50
1	1	100	3.00	1.50
2	1	100	3.00	1.50
3	1	100	3.00	1.50
1	3	0	2.00	—
2	2	0	3.00	—
3	1	0	6.00	—
1	3	2	2.05	1.21
2	2	2	3.08	1.21
3	1	1	6.16	2.42

multiplier for a given horse is always higher than the placing multiplier for that same horse.

It's easy to extend this system to cover winning, placing, and showing multipliers, and also trifectas (predicting exactly which horse will win, which will place, which will show) and other betting options. At no time is the track owner in risk of losing money. If the track owners choose to take a fixed amount (subtract a fixed amount rather than calculate a percentage) of the bets placed, as long as enough people are showing up and betting, then the track owners know to the penny how much they'll bring in every night.

Horse racing is interesting because bettors can decide for themselves whether this is gambling or a game of skill. From the track owners' point of view, the debate is moot. Their expected value of return is positive as long as people show up and place bets. This is very different from if someone really could predict the results of a roulette wheel spin (or just kept betting on the correct results by dumb luck). The roulette wheel winner cannot afford a significant long-term deviation from the statistical predictions. The race track owner doesn't care.

What about expected value of return in a pari-mutuel betting game? In order to calculate an expected value, I need to know the probability of a horse winning. I don't know this; I only know the perceived probability based on the bets placed. If everyone betting is truly an expert on the horses, then everybody would bet on the horse that's about to win and everyone would lose money. If there are indeed true experts, then they can only win if there are also a bunch of people betting who aren't experts at all.

Finally, what about games such as poker? If you are really the most skillful player at the table, then in the long run, you will win. Staying in the game for the long run, however, could involve a lot of hands played, which of course means that you might need very deep pockets to cover your losses and keep you in the game long enough for your skills to prevail.

PROBLEMS

The problems for Chapter 14 are just for fun. There's nothing to learn about games of chance other than that you can't beat the house. Games like poker combine chance, actual skill at playing hands, and psychological effects such as bluffing; they're very hard to quantify.

1. The coin flip game described in this chapter is mathematically equivalent to a "one-dimensional random walk." What this means is that you flip the coin; if you get heads, you take a step forward; if you get tails, you take an equal-sized step backward. The number of steps forward you get is equivalent to dollars won; the number of steps backward is equivalent to dollars lost. If you start at position +1, it's equivalent to starting with $1 in your pocket; if you start at +2, it's equivalent to starting with $2 in your pocket; and so on.

   Try it. You can either actually flip a coin and take steps or you can do it on a spreadsheet. Most spreadsheet programs have a random number generator that gives you a random number between 0 and 1. Generate a column of, say, 100 of these. Then in the next column, use an if statement: If your random number is >0.5, put +1; otherwise, put −1. In the third column, keep a running total of all the numbers in the second column.

   Start the column with a number which is the amount of money you started the game with. Show how long you can play without getting wiped out.

2. A state lottery offers a prize of $400 million. A single ticket's probability of winning is 1/400,000,000. Assume that the state wants to make $100 million for holding the lottery and that tickets cost $1 a piece.

   (a) What's the expected value of a lottery ticket?

   (b) Just to help get a feeling for the odds here, about how long (seconds in years) is 1 second in 500 million seconds?

Chapter 15

Spreadsheet Calculators

15.1 INTRODUCTION TO THE SPREADSHEETS

Two approaches to providing readers with computer-based calculators are employed in this book. One approach is to provide links to the many online calculators that provide slick interfaces and useful tutorials. A second approach is to create spreadsheets that, while not quite as elegant looking as the online calculators, allow the user to examine the calculations, modify and/or add features if desired, make available some intermediate or extended results, generate reports, and so on. Also, by creating a set of custom spreadsheets tied to this book, I can provide a consistent interface among all of them.

An advantage of the online calculators for me is that I don't have to create them; I just find them and then provide links at the appropriate places in the book. The disadvantage is that I cannot guarantee that these links will be supported and will be there when you want them. I've tried to play the odds by providing several links for every calculator. If they've all disappeared when you want them, you'll have to go searching for new ones.

In order to give readers access to my spreadsheets, I am posting them on my website (www.lawrencedworsky.com). All you have to do is to click on the desired spreadsheet link and there it is. You will need a spreadsheet program on your computer that can read Microsoft Office® Excel® spreadsheets.[1] The Excel format is as universal a format as you can find; almost every modern spreadsheet program will read it. When you click on one of these spreadsheets, you will be given the choice of opening it with your spreadsheet program or saving it to some location on your hard disk. If you choose to open it with your spreadsheet program, you can then save it to your hard disk at a later time. In either case, I recommend saving my original spreadsheet to your hard disk and then working with a copy of it. Also, you

---

[1] I created these spreadsheets using the 1997–2003 Excel file format (.xls). This format will work on any version of Excel sold in the past 12 years or so, including the current version, as well as most other spreadsheet programs.

---

should check back with the website occasionally to see if there have been any updates to any of these spreadsheets.

In order to use these spreadsheets, you must have a basic familiarity with how to use a modern spreadsheet. Your ability to perform basic operations such as scrolling, entering data, cutting or copying and pasting, printing a sheet or a part of a sheet, generating a simple formula in a cell, and so on is required. Since I don't know which spreadsheet program you'll be using, it's not a good idea for me to try to provide a primer. There are very many good introductory books on spreadsheet use available.

I have deliberately not protected or locked any of the cells in these spreadsheets. This means that these spreadsheets can be modified to any extent, including to the extent of corrupting or deleting the calculations. In other words, make sure that you haven't changed something and then forgotten about your change. You can always download the original spreadsheet again if you have to; you can't modify the copy stored on my website. I recommend saving any modified versions under new names so you can reuse the original without losing your changes.

In many cases, I have taken somewhat circuitous approaches to writing these spreadsheets rather than using the many elegant and concise formulas available in the spreadsheet programs themselves. The results would be the same whichever way I had done it, but by actually working through the calculations rather than using the formulas, I can tie the spreadsheet calculations to the text materials and then you can follow what's happening as closely as you want.

Most spreadsheets have excellent graphic capabilities. I recommend that you familiarize yourself with these capabilities as you use the spreadsheets. Get into the habit of graphing results as often as you can. When you're doing comparisons, try to save intermediate results to a blank spreadsheet and then graph the comparisons. Once you develop the skill to create graphs quickly, you'll really come to appreciate the power of "seeing things" that spreadsheets bring to you.

Lastly, since you are free to modify these spreadsheets as you wish and I of course have no idea just what modifications you might come up with, I don't error-check input data. For example, when the spreadsheet asks for a month, you should enter a number between 1 and 12. Anything else will cause generally unpredictable and almost definitely useless results. On the other hand, you can't damage either the hardware or the software by entering nonsense. Don't be afraid to try things—it's the best way to learn.

If you don't have a spreadsheet program and would like a free one, I recommend the following packages. These packages are all full office suites, giving you a word processor, a presentation program, possibly some other programs, and of course a spreadsheet:

1. Open Office (www.openoffice.org for the download);
2. IBM Lotus Symphony (http://symphony.lotus.com/software/lotus/ symphony/home.nsf/products for the download);
3. Google Docs (www.google.com); choose "documents" from the "more" tab.

	A	B	C	D	E	F	G	H	I	J	K	L
1	**Start Month:**	5		**Pmt Nr**	**Mnth**	**Year**		**Balance**	**Payment**	**Interest**	**Tot Int / Year**	
2	**Start Year:**	2008		0	5	2008		$65,000.00	$0.00	$0.00	$0.00	
3				1	6	2008		$61,589.21	$3,844.12	$433.33	$433.33	
4				2	7	2008		$58,155.69	$3,844.12	$410.59	$843.93	
5	**Nr Mnthly Pmts**	18		3	8	2008		$54,699.27	$3,844.12	$387.70	$1,231.63	
6				4	9	2008		$51,219.82	$3,844.12	$364.66	$1,596.29	
7	**Principal**	$65,000		5	10	2008		$47,717.16	$3,844.12	$341.47	$1,937.76	
8	**Rate**	8.00%		6	11	2008		$44,191.16	$3,844.12	$318.11	$2,255.87	
9				7	12	2008		$40,641.65	$3,844.12	$294.61	$2,550.48	
10				8	1	2009		$37,068.47	$3,844.12	$270.94	$270.94	
11				9	2	2009		$33,471.47	$3,844.12	$247.12	$518.07	
12				10	3	2009		$29,850.50	$3,844.12	$223.14	$741.21	
13				11	4	2009		$26,205.38	$3,844.12	$199.00	$940.21	
14				12	5	2009		$22,535.97	$3,844.12	$174.70	$1,114.92	
15				13	6	2009		$18,842.09	$3,844.12	$150.24	$1,265.16	
16				14	7	2009		$15,123.58	$3,844.12	$125.61	$1,390.77	
17				15	8	2009		$11,380.29	$3,844.12	$100.82	$1,491.59	
18				16	9	2009		$7,612.03	$3,844.12	$75.87	$1,567.46	
19				17	10	2009		$3,818.66	$3,844.12	$50.75	$1,618.21	
20				18	11	2009		$0.00	$3,844.12	$25.46	$1,643.67	
21												
22												

H ◄ ► H   **Loan**   Save   Loan V2

Figure 15.1

The first two of these are conventional software packages in that you download the package and install it on your computer, pretty much the same as if you had bought the software on a disk. The last is an online software—everything is stored online. Software has recently become available to let you take your work offline.

All of these spreadsheets have a similar "look and feel." Figure 15.1 is a picture of the worksheet portion of the spreadsheet that accompanies Chapter 3. Its file name is Ch3Amortization.xls. Don't worry about what the calculations actually mean until you're reading Chapter 3. My purpose here is just to show you how to get around the spreadsheet. Different spreadsheet programs might look a bit different from this, particularly outside the actual work area.

If you can download this spreadsheet and open it in your spreadsheet program, I recommend doing so now. If not, look at the figures of this chapter but keep in mind that when you do open this spreadsheet with your software, it might look a little different.

Near the bottom of the spreadsheet, on the left, you'll see the terms Loan, Save, and Loan V2. Each of these terms is located in a tab. These tabs are actually separate spreadsheets that are all relevant to the chapter and that I created and saved as a single file.

On the left side of the sheet is a vertical green line. If you scroll down, you'll see that this line extends all the way down to row 1,000. I've set up all of the sheets to automatically handle situations with up to 1,000 line items. For example, if you are studying a loan with monthly payments, 1,000 line items is about 83 years. I think is a big enough capability. You can easily extend this if necessary.

To the left of the green line is a set of labels (first column) and numbers accompanying these labels (second column). I refer to the area to the left of the green line as the input area. All, or in some cases that I'll get into later, almost all, of the information needed to set up your problem is entered in the input area. The input area formats all information properly. For example, if you enter 0.0773 for the Rate, when you hit Enter, the number you just typed changes from 0.0773 to 7.73%.

Let me repeat that there is no error checking in these spreadsheets. If you enter 23 for the Start Month and −6 for the Number of Monthly Payments (Nr Mnthly Pmts), I can't predict what will happen, but I'm pretty sure that it will be worthless.

As long as you're only entering numbers in the input area, no permanent changes are being made to the structure of the spreadsheet. Because of this, you can change these numbers to your heart's content. When you repeat a set of input numbers, you should expect the exact same output numbers to appear.

Try changing the Nr Mnthly Pmts to anything from 2 to 1,000. Not only will all the numbers on the sheet adjust, but the sheet will start at Payment Number (Pmt Nr) 0 (just to the right of the green line) and terminate at the number you just entered. You don't have to worry about how I did this until you want to print your spreadsheet. I'll discuss this below.

To the right of the green line is the area that's used for output and also for capturing some inputs that can't be easily summarized on the left of the green line.

The first three columns to the right of the green line, in this spreadsheet, are the Payment Number (Pmt Nr), the Month (Mnth), and the Year. This particular spreadsheet is used for studying loans such as mortgage loans. The first month and year shown are the month and year that you took the loan. Since the first payment is due 1 month later, I refer to the day you took the loan as Pmt Nr 0.

Every column to the right of the Pmt Nr column refers to the Pmt Nr column, as is standard for tables. For example, after Pmt Nr 12, which takes place in May of 2009, the balance is $22,535.97, the payment was $3,844.12, the interest paid was $174.40, and the total interest paid in 2009 was $1,114.92.

Note that the payment is the same for all 18 payments (this is the default situation in a fixed interest, fixed payment mortgage) and that this payment amount was supplied by the spreadsheet.

The balance column entries get smaller each month, ending up at 0 after the eighteenth payment is made. This is exactly what the payment amount was calculated to accomplish.

The total interest per year column is for tax reporting purposes, used when you have a loan tax deductible interest. This number starts low each January and grows through December. If what's happening isn't clear to you, change the Nr Mnthly Pmts to, say, 60 (for a 5-year loan) and look at the total interest per year column again.

If you put the cursor in one of the Payment cells other than the top one and look just above the actual work area columns, you'll see a pretty messy-looking expression, for example,

$$=IF(D14<>"",-PMT(\$B\$8/12,nr-D14+1,H13+J14,0,1),"").$$

This expression is doing the work of calculating the monthly payment. Don't worry about the IF part; I'll get to that below. The actual calculating engine is the internal spreadsheet function:

$$PMT(\$B\$8/12,nr-D14+1,H13+J14,0,1).$$

If you don't understand what's happening here, I recommend a book on spreadsheet writing or the internal Help system of your spreadsheet program. I use this same internal function every time I want to calculate the month's payment, based on the interest rate, the number of payments to go, the current balance, and this month's interest.

It would have been simpler to calculate the regular payment just once and then just copy this number down the column. The reason I did things the hard way is that this approach makes the spreadsheet itself dynamic in that if one of the payments is changed manually, the spreadsheet corrects itself automatically.

For example, suppose I make a $10,000 payment for Pmt Nr 9. If you have the spreadsheet running, try just entering $10,000 into the appropriate payment cell. Your spreadsheet should now look like Figure 15.2. If you don't have the spreadsheet running, just look at Figure 15.2.

The spreadsheet has recalculated a new regular payment to bring your balance to 0 after 18 months, has recalculated the correct balance after your payment, and so on. My purpose here is not to repeat the explanations of Chapter 3 but to discuss

	A	B	C	D	E	F	G	H	I	J	K	L	M	N
1	Start Month:	5		Pmt Nr	Mnth	Year		Balance	Payment	Interest	Tot Int / Year			
2	Start Year:	2008		0	5	2008		$65,000.00	$0.00	$0.00	$0.00			
3				1	6	2008		$61,589.21	$3,844.12	$433.33	$433.33			
4				2	7	2008		$58,155.69	$3,844.12	$410.59	$843.93			
5	Nr Mnthly Pmts	18		3	8	2008		$54,699.27	$3,844.12	$387.70	$1,231.63			
6				4	9	2008		$51,219.82	$3,844.12	$364.66	$1,596.29			
7	Principal	$65,000		5	10	2008		$47,717.16	$3,844.12	$341.47	$1,937.76			
8	Rate	8.00%		6	11	2008		$44,191.16	$3,844.12	$318.11	$2,255.87			
9				7	12	2008		$40,641.65	$3,844.12	$294.61	$2,550.48			
10				8	1	2009		$37,068.47	$3,844.12	$270.94	$270.94			
11				9	2	2009		$27,315.59	$10,000.00	$247.12	$518.07			
12				10	3	2009		$24,360.57	$3,137.13	$182.10	$700.17			
13				11	4	2009		$21,385.84	$3,137.13	$162.40	$862.58			
14				12	5	2009		$18,391.28	$3,137.13	$142.57	$1,005.15			
15				13	6	2009		$15,376.76	$3,137.13	$122.61	$1,127.76			
16				14	7	2009		$12,342.14	$3,137.13	$102.51	$1,230.27			
17				15	8	2009		$9,287.29	$3,137.13	$82.28	$1,312.55			
18				16	9	2009		$6,212.07	$3,137.13	$61.92	$1,374.46			
19				17	10	2009		$3,116.36	$3,137.13	$41.41	$1,415.88			
20				18	11	2009		$0.00	$3,137.13	$20.78	$1,436.65			
21														
22														
23														
24														
25														
26														

Loan   Save   Loan V2

Figure 15.2

about the spreadsheet operation, so I won't discuss the meanings of these changes to your loan.

What if I want to examine the results of some other changes? For example, what if I want to look at a loan with a different principal, or what if I want to move this big payment from Pmt Nr 9 to Pmt Nr 7? If I just put in these new changes, suddenly I have a mess on my hands. The Pmt Nr 9 cell no longer has its original formula; it's frozen at $10,000 regardless of whether or not this number makes sense together with the new numbers. Also, every result on the sheet below the $10,000 entry is suspect because the rest of the spreadsheet doesn't "know" that this number might be bogus.

Once something has been changed to the right of the green line, the spreadsheet has to be reinitialized before it is used again. This can be done in one of three ways:

1. Close the spreadsheet (save it to another name first if you want to keep it). Then reload the original spreadsheet either from my website or from a copy you've made of the original spreadsheet. This is the safest way of ensuring that you've got all the original formulas back. It can be annoying, however, because you have to reenter all of your information to the left of the green line.

2. Every time you press the key combination Ctrl-z (hold the control key down and press the z key), the last entry you made in the spreadsheet is undone. Equivalently, your spreadsheet program might have Undo and Redo icons on the page somewhere. Just keep track of how many times you Undo (and/ or Redo) and examine everything carefully to make sure you didn't undo one change too many or too few.

3. Go to another cell on the spreadsheet that has the same formula (in the above example another of the Payment cells) and type Ctrl-C (copy the formula). Then go to the cell that has the $10,000 entry and type Ctrl-V (paste the formula). This works fine, but you have to keep careful track of what you're copying from where and to where. This approach is most useful when you have to undo a range of changes, for example, if you changed almost every Payment in the spreadsheet one at a time and you want to undo all of this at once. This will even work if you've changed every Payment from number 1 to number 18—see the programming notes below.

The different tabs in a spreadsheet will bring up different sheets that are related by the chapter in the book that this spreadsheet "belongs to." The tab system has the ability of tying information from one sheet to another in a given spreadsheet file, but I'm not using this capability. I'm only using the tab capability to help me orga-nize the spreadsheets to match the chapters in the book. Each tab represents (and could have been written as) a completely independent spreadsheet file. I did this in anticipation of user modifications of the spreadsheets; by keeping each spreadsheet independent of all other spreadsheets, there is no chance of inadvertent changes to a spreadsheet that you're not looking at.

	A	B	C	D	E	F	G	H	I	J	K	L	M	N
1	Start Month:	1		Pmt Nr	Month	Year		Rate	Balance	Payment	Interest	Tot Int / Year		
2	Start Year:	2010		0	1	2007		6.18%	$325,000.00	$0.00	$0.00	$0.00		
3	Nr Pmts	240		1	2	2007		6.18%	$325,000.00	$1,673.75	$1,673.75	$1,673.75		
4				2	3	2007		6.18%	$325,000.00	$1,673.75	$1,673.75	$3,347.50		
5	Principal	$325,000		3	4	2007		6.18%	$325,000.00	$1,673.75	$1,673.75	$5,021.25		
6				4	5	2007		6.18%	$325,000.00	$1,673.75	$1,673.75	$6,695.00		
7				5	6	2007		6.18%	$325,000.00	$1,673.75	$1,673.75	$8,368.75		
8				6	7	2007		6.18%	$325,000.00	$1,673.75	$1,673.75	$10,042.50		
9				7	8	2007		6.18%	$325,000.00	$1,673.75	$1,673.75	$11,716.25		
10				8	9	2007		6.18%	$325,000.00	$1,673.75	$1,673.75	$13,390.00		
11				9	10	2007		6.18%	$325,000.00	$1,673.75	$1,673.75	$15,063.75		
12				10	11	2007		6.18%	$325,000.00	$1,673.75	$1,673.75	$16,737.50		
13				11	12	2007		6.18%	$325,000.00	$1,673.75	$1,673.75	$18,411.25		
14				12	1	2008		6.18%	$325,000.00	$1,673.75	$1,673.75	$1,673.75		
15				13	2	2008		6.18%	$325,000.00	$1,673.75	$1,673.75	$3,347.50		
16				14	3	2008		6.18%	$325,000.00	$1,673.75	$1,673.75	$5,021.25		
17				15	4	2008		6.18%	$325,000.00	$1,673.75	$1,673.75	$6,695.00		
18				16	5	2008		6.18%	$325,000.00	$1,673.75	$1,673.75	$8,368.75		
19				17	6	2008		6.18%	$325,000.00	$1,673.75	$1,673.75	$10,042.50		
20		·		18	7	2008		6.18%	$325,000.00	$1,673.75	$1,673.75	$11,716.25		
21				19	8	2008		6.18%	$325,000.00	$1,673.75	$1,673.75	$13,390.00		
22				20	9	2008		6.18%	$325,000.00	$1,673.75	$1,673.75	$15,063.75		
23				21	10	2008		6.18%	$325,000.00	$1,673.75	$1,673.75	$16,737.50		
24				22	11	2008		6.18%	$325,000.00	$1,673.75	$1,673.75	$18,411.25		
25				23	12	2008		6.18%	$325,000.00	$1,673.75	$1,673.75	$20,085.00		
26				24	1	2009		6.18%	$325,000.00	$1,673.75	$1,673.75	$1,673.75		

Basic **ARM**

Figure 15.3

The Basic tab of the spreadsheet Ch4Mortgages.xls is identical to the Basic tab of the spreadsheet Ch3Amortization.xls. Paying off a fixed rate, fixed payment mortgage is identically a loan amortization.

Figure 15.3 shows part of the spreadsheet on the ARM tab of the Ch4Mortgages. xls. This spreadsheet is the same as the Basic tab spreadsheet except that the Rate variable has disappeared from the left side of the green line and has appeared as a column to the right of the green line.

ARM stands for adjustable rate mortgage. This spreadsheet has been prepared to accommodate interest rates that may change once or even many times over the course of the mortgage. To use this spreadsheet, you first enter all the required information to the left of the green line, then you enter the initial interest rate in the top cell of the Rate column. The number you enter automatically gets copied down to the bottom of the sheet (6.18% in Figure 15.4).

Suppose that the interest changes to 7.00% after a year (I don't think this is a legal change for mortgages in most states, but I'm doing it just to keep my example contained in one page). "After 1 year" in this example is January of 2008, or equivalently at Pmt Nr 12. What you need to do is to enter 7% (0.07) into the Rate cell at Pmt Nr 12. The sheet automatically changes all the rates from Pmt Nr 12 down to the bottom of the sheet to 7%. At Pmt Nr 16, change the rate again, this time down to 6.50%. Enter 6.50% at the appropriate cell and all the rates from Pmt Nr 16 down to the bottom of the sheet change to this latest rate.

Figure 15.4 is part of the ARM spreadsheet showing these changes. The balance, payment, and so on columns are all automatically correctly updated. In Figure 15.4,

	A	B	C	D	E	F	G	H	I	J	K	L	M	N
1	Start Month:	1		Pmt Nr	Month	Year		Rate	Balance	Payment	Interest	Tot Int / Year		
2	Start Year:	2010		0	1	2007		6.18%	$325,000.00	$0.00	$0.00	$0.00		
3	Nr Pmts	240		1	2	2007		6.18%	$324,311.47	$2,362.28	$1,673.75	$1,673.75		
4				2	3	2007		6.18%	$323,619.40	$2,362.28	$1,670.20	$3,343.95		
5	Principal	$325,000		3	4	2007		6.18%	$322,923.77	$2,362.28	$1,666.64	$5,010.59		
6				4	5	2007		6.18%	$322,224.55	$2,362.28	$1,663.06	$6,673.65		
7				5	6	2007		6.18%	$321,521.73	$2,362.28	$1,659.46	$8,333.11		
8				6	7	2007		6.18%	$320,815.29	$2,362.28	$1,655.84	$9,988.94		
9				7	8	2007		6.18%	$320,105.21	$2,362.28	$1,652.20	$11,641.14		
10				8	9	2007		6.18%	$319,391.48	$2,362.28	$1,648.54	$13,289.69		
11				9	10	2007		6.18%	$318,674.07	$2,362.28	$1,644.87	$14,934.55		
12				10	11	2007		6.18%	$317,952.97	$2,362.28	$1,641.17	$16,575.72		
13				11	12	2007		7.00%	$317,228.15	$2,362.28	$1,637.46	$18,213.18		
14				12	1	2008		7.00%	$316,564.52	$2,514.13	$1,850.50	$1,850.50		
15				13	2	2008		7.00%	$315,897.01	$2,514.13	$1,846.63	$3,697.12		
16				14	3	2008		7.00%	$307,739.74	$10,000.00	$1,842.73	$5,539.86		
17				15	4	2008		7.00%	$307,080.47	$2,454.43	$1,795.15	$7,335.00		
18				16	5	2008		6.50%	$306,417.34	$2,454.43	$1,791.30	$9,126.31		
19				17	6	2008		6.50%	$305,712.17	$2,364.94	$1,659.76	$10,786.07		
20				18	7	2008		6.50%	$305,003.17	$2,364.94	$1,655.94	$12,442.01		
21				19	8	2008		6.50%	$304,290.33	$2,364.94	$1,652.10	$14,094.11		
22				20	9	2008		6.50%	$303,573.64	$2,364.94	$1,648.24	$15,742.35		
23				21	10	2008		6.50%	$302,853.06	$2,364.94	$1,644.36	$17,386.71		
24				22	11	2008		6.50%	$302,128.57	$2,364.94	$1,640.45	$19,027.16		
25				23	12	2008		6.50%	$301,400.17	$2,364.94	$1,636.53	$20,663.69		
26				24	1	2009		6.50%	$300,667.81	$2,364.94	$1,632.58	$1,632.58		

Basic ARM

Figure 15.4

I have also changed Pmt Nr 14 to $10,000 just to show that this type of change will also be correctly handled by the spreadsheet.

15.2 SOME PROGRAMMING NOTES

When you enter a number of payments in one of these spreadsheets, the sheet shows entries from Pmt Nr 0 or 1, as appropriate for the topic at hand, up to the number of payments entered. The actual spreadsheet, however, has been programmed down to row 1,000. If you click on a blank cell in the Payment column of the spreadsheets I used above, you'll see the same formula as in the cells with visible data. If you need to recover the formula in the payment column (or any other column), you can easily do it by copying one of these "blank" cells up into the required cells above it.

The sheets' ability to stop showing anything after the specified number of payments has been reached is handled by the "If" statements that are cluttering up all the cell formulas. The Pmt Nr column formula is simple; add 1 to the previous cell (the cell above) and check to see if the result is larger than the Nr Mnthly Pmts. If it isn't larger, show the result. If it is larger, don't show anything.

Every other cell to the right of the Pmt Nr column looks left to the Pmt Nr cell. If the Pmt Nr cell is showing a number, perform the specified calculation and show it. If the Pmt Nr cell is not showing anything, don't show anything.

If you want to print your results using these spreadsheets, you have to do a little extra work. Spreadsheet programs automatically check to see which cells have entries and then designate a print area to include these cells. What you need to do is to reset the print area specifically to show only the region of the sheet where you have useful information. This is very easy to do—but you will have to find out how to do it with your particular spreadsheet program. If you don't do it, you'll print not only the material you want but also the many blank pages representing all the rows down to row 1,000.

I've set the formatting of the cells in these spreadsheets so that percentages are shown to two decimal places. Amounts of money are usually shown to the penny except where this level of resolution probably isn't necessary. For example, in Figures 15.1 and 15.2, you'll see balances, payments, and interest shown to the penny but the principal is only shown to the nearest dollar. This can all be changed with the formatting available in the spreadsheet program. Remember that the spreadsheet is internally carrying all numbers to a high level of resolution and just rounding this internal resolution to the presentation resolution you've specified. The accuracy of your results does not depend on your format. Suppose you bought three items that each cost $33.33. Your total cost is exactly $99.99. If, however, you only display numbers to the nearest dollar, on the spreadsheet it would look as if you were adding up three $33 dollar items and getting $100.

Lastly, I plan to upgrade the online spreadsheets and add capabilities regularly. If I find a bug or one is reported to me, I'll list the bug on the Errata page and correct the spreadsheet. All corrections and upgrades will be documented on my website.

Chapter 16

Solutions

16.1 CHAPTER 1

1.

    (a) $7 - (12 - 5) = 7 - 7 = 0$

    (b) $12(14 - 6) = 12(8) = 96$

    (c) $\dfrac{16-(3+7)}{3(7-5)} = \dfrac{16-10}{3(2)} = \dfrac{6}{6} = 1$

    (d) $(12 - 2)(7 + 3) = (10)(10) = 100$

    (e) $12 - 2(7 + 3) = 12 - 2(10) = 12 - 20 = -8$

    (f) $(12 - 2)7 + 3 = (10)7 + 3 = 70 + 3 = 73$

    (g) $6.2 + 1/3 = 6.2 + 0.333 = 6.533 \sim 6.53$

2.

    (a) $x + y + z = 6 + 2 + 3 = 9$

    (b) $z(x - 3)(y + 2) = 3(6 - 3)(2 + 2) = 6(3)(4) = 72$

    (c) $\dfrac{\frac{x+2}{y-3}}{z-4} + 2.25 = \dfrac{\frac{6+2}{2-3}}{3-4} + 2.25 = \dfrac{\frac{8}{-1}}{-1} + 2.25 = \dfrac{-8}{-1} + 2.25 = 8 + 2.25 = 10.25$

    (d) $x(x - 1)(x + 2) = 6(6 - 1)(6 + 2) = 6(5)(8) = 240$

3.

    (a) $T_1 = 0$
          $P_3 = 0$
          $N_4 = 4$
          $T_5 = 8$

Understanding the Mathematics of Personal Finance: An Introduction to Financial Literacy, by
Lawrence N. Dworsky
Copyright © 2009 John Wiley & Sons, Inc.

(b) At more than \$20 per wallet, $\sum_{i=1}^{4} N_i = 6 + 2 + 2 + 4 = 14$.

All together, $\sum_{i=1}^{6} N_i = 6 + 2 + 2 + 4 + 8 + 16 = 38..$

(c) Total cost of wallets sold is 38(\$12) = \$456.

Total revenues is $\sum_{i=1}^{6} N_i P_i = 6(25) + 2(24) + 2(23) + 4(18) + 16(14) = 696$.

(Total revenues) − (Cost of wallets sold) = \$696 − \$456 = \$240.

Subtracting overhead, \$240 − \$100 = \$140. Money was made for the day.

4.

(a)

T (hours)	P (\$)	N (number of wallets sold)
0	22.50	9
2	21.60	3
4	20.70	3
6	18.90	6
8	16.40	12
10	12.60	24

(b) The revenue for the second day was \$939.60. The storekeeper's cost for the wallets was 57(\$12) = \$684.00. The total profit for the day, including overhead, was then \$939.60 − \$684.00 − \$100.00 = \$155.60.

5. See Figure 16.1.

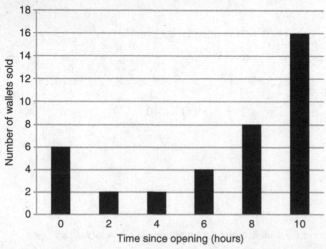

Figure 16.1

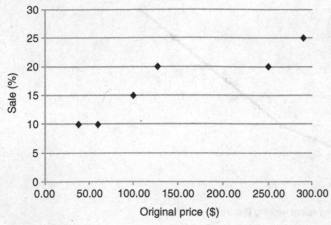

Figure 16.2

6.

Original price ($)	Sale price ($)	% Change	% Sale
289.99	217.49	75	25
249.99	199.99	80	20
127.50	102.00	80	20
99.99	84.99	85	15
59.79	53.99	90	10
37.50	33.75	90	10

See Figure 16.2.

7.

n	$(0.25 + x)^n$	$(0.25 + x)^n$
	$x = 0.5$	$x = 1.2$
0	1	1
1	0.75	1.45
2	0.57	2.10
3	0.42	3.05
4	0.32	6.41

A number raised to the 0th power is always equal to 1.
A number raised to the 1st power is always equal to the number itself.

8. $13, $22, $54, and $1,720

9. See Figure 16.3.
The average speed is the total distance divided by the total time: 3.0/2.0 = 1.5 mph.
This walk was taken at two steady speeds (two straight line segments on the graph).

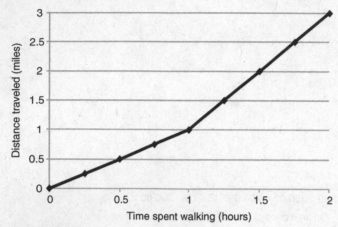

Figure 16.3

For the first hour, 1 mi was traveled. The instantaneous speed any time in the first hour was therefore $1/1 = 1$ mph. For the second hour, $(3 - 1) = 2$ mi was traveled. The instantaneous speed any time in the second hour was therefore $2/1 = 2$ mph.[1]

16.2 CHAPTER 2

1.

(a) Interest = 10%($6,700.00) = 0.1($6,700.00) = $670.00

(b) Interest = 3(0.06)($500.00) = $90.00

(c) Interest = $\frac{7}{12}$(0.08)($1,000) = $46.67

(d) Interest = $15,600.00 − $12,000.00 = $3,600.00
Interest per year = $3,600.00/3 = $1,200.00
Interest rate = $1,200.00/12,000.00 = 12/120 = 1/10 = 0.1 = 10%

2. 18.00%, 9.00%, 6.00%, and 4.50%

3. By hand (or pocket calculator):

After 1 year, $1,250.00 + (0.075)1,250.00 = 1.075(1,250.00) = $1,343.75.

After 2 years, 1.075(1,343.75) = $1,444.53.

After 3 years, 1.075(1,444.53) = $1,552.87.

Alternate approach (by pocket calculator):

After 1 year, ($1,250)(1.075)^1 = $1,343.75.

After 2 years, ($1,250)(1.075)^2 = $1,444.53.

After 3 years, ($1,250)(1.075)^3 = $1,552.87.

---

[1] At exactly 1 hour, the slope of the graph changes abruptly. Rigorously speaking, we cannot define an instantaneous speed at exactly this time. In the terminology of differential calculus, the derivative is discontinuous at Time = 1.00 and does not have a unique value.

Using the spreadsheet:

		Cmpdng interval	Interest ($)	Balance ($)
Nr Years	3	0	0.00	1,250.00
Cmpds per Year	1	1	93.75	1,343.75
		2	100.78	1,444.53
Principal	$1,250.00	3	108.34	1,552.87
Rate	7.50%			
Init	$0.00			

4. Using the spreadsheet:

		Cmpdng interval	Interest ($)	Balance ($)
Nr Years	1.5	0	0.00	10,000.00
Cmpds per Year	12	1	62.50	10,062.50
		2	62.89	10,125.39
Principal	$10,000.00	3	63.28	10,188.67
Rate	7.50%	4	63.68	10,252.35
		5	64.08	10,316.43
Init	$0.00	6	64.48	10,380.91
		7	64.88	10,445.79
		8	65.29	10,511.08
		9	65.69	10,576.77
		10	66.10	10,642.87
		11	66.52	10,709.39
		12	66.93	10,776.33
		13	67.35	10,843.68
		14	67.77	10,911.45
		15	68.20	10,979.65
		16	68.62	11,048.27
		17	69.05	11,117.32
		18	69.48	11,186.81

5. Using the spreadsheet:

			Rate per day:	0.0223%	
		Cmpdng interval	Interest ($)	Balance ($)	Payoff
Nr Years	15	0	0.00	10,000.00	
Cmpds per Year	12	1	62.50	10,062.50	
		2	62.89	10,125.39	
Principal	$1,000.00	3	63.28	10,188.67	

Continued

			Rate per day:	0.0223%	
		Cmpdng interval	Interest ($)	Balance ($)	Payoff
Rate	7.50%	4	63.68	10,252.35	
		5	64.08	10,316.43	
Init	$0.00	6	64.48	10,380.91	
		7	64.88	10,445.79	
Proration		8	65.29	10,511.08	
Nr of days in month	28	9	65.69	10,576.77	
		10	66.10	10,642.87	
Day of proration	13	11	66.52	10,709.39	
Payoff Month Nr	11				$10,740.47

6. Using the spreadsheet:

		Cmpdng interval	Interest ($)	Balance ($)
Nr Years	1.5	0	0.00	10,250.00
Cmpds per Year	12	1	64.06	10,314.06
		2	64.46	10,378.53
Principal	$10,000.00	3	64.87	10,443.39
Rate	7.50%	4	65.27	10,508.66
		5	65.68	10,574.34
Init	$250.00	6	66.09	10,640.43
		7	66.50	10,706.93
		8	66.92	10,773.85
		9	67.34	10,841.19
		10	67.76	10,908.95
		11	68.18	10,977.13
		12	68.61	11,045.73
		13	69.04	11,114.77
		14	69.47	11,184.24
		15	69.90	11,254.14
		16	70.34	11,324.48
		17	70.78	11,395.26
		18	71.22	11,466.48

The balance at the end of 18 months has increased from $11,186.81 to $11,466.48. Going back to the spreadsheet, I will return Init (the up-front cost) to 0 and then tweak the Rate until I again get this new balance (or as close as I can by changing the interest rate to two decimal places):

		Cmpdng interval	Interest ($)	Balance ($)
Nr Years	1.5	0	0.00	10,000.00
Cmpds per Year	12	1	76.33	10,076.33
		2	76.92	10,153.25
Principal	$10,000.00	3	77.50	10,230.75
Rate	9.16%	4	78.09	10,308.85
		5	78.69	10,387.54
Init	$0.00	6	79.29	10,466.83
		7	79.90	10,546.73
		8	80.51	10,627.23
		9	81.12	10,708.35
		10	81.74	10,790.09
		11	82.36	10,872.46
		12	82.99	10,955.45
		13	83.63	11,039.08
		14	84.26	11,123.34
		15	84.91	11,208.25
		16	85.56	11,293.81
		17	86.21	11,380.02
		18	86.87	11,466.89

7.

Loan rate (%)	10-year balance ($)	Savings rate (%)	10-year balance ($)	Profit ($)
5	16,289	6	17,908	1,619
5	16,289	7	19,672	3,383
5	16,289	8	21,589	5,300
6	17,908	7	19,672	1,764
6	17,908	8	21,589	3,681
7	19,672	8	21,589	1,917

16.3 CHAPTER 3

1.

(a) $843.86

(b) $2,048.18

(c) $332.14

(d) Look at solution 1a above. This is the same loan except for the principal. Therefore, scaling will work. The amount $250,000 is 2.5($100,000), so the payment is 2.5($843.86) = $2,109.65.

2. 2005 $2,482.72
    2010 $4,601.21
    2015 $2,673.79
    2020 $116.39

3. After payment number 110, in September of 2014.

4. May of 2013 is payment number 94. Replacing the regular payment of $843.86 with $10,000, the regular payment for the rest of the loan falls to $712.60:

Pmt Nr	Mnth	Year	Balance ($)	Payment ($)	Interest ($)	Tot Int/Year ($)
92	3	2013	59,957.23	843.86	302.49	915.55
93	4	2013	59,413.16	843.86	299.79	1,215.33
94	5	2013	49,710.22	10,000.00	297.07	1,512.40
95	6	2013	49,246.17	712.60	248.55	1,760.95
96	7	2013	48,779.80	712.60	246.23	2,007.18

5. This requires a little planning. For the amortization table from problem 2, note that the balance after payment number 3 is $98,963.26, after payment number 6 is $97,910.10, and so on. The first month's interest is $500.00, so change payment number 1 to $500.00. Again, everything readjusts and the second month's interest is now $500. Set the second month's payment to $500.

Now "tinker with" the third payment until the balance is again $98,963.26. You'll find that the third payment must be $1,536.74. Continuing, at the fourth payment, the interest is $494.82, so set the fourth payment to $494.82 and so on:

Pmt Nr	Mnth	Year	Balance ($)	Payment ($)	Interest ($)	Tot Int/Year ($)
0	1	2005	100,000.00	0.00	0.00	0.00
1	2	2005	100,000.00	500.00	500.00	500.00
2	3	2005	100,000.00	500.00	500.00	1,000.00
3	4	2005	98,963.26	1,536.74	500.00	1,500.00
4	5	2005	98,963.26	494.82	494.82	1,994.82
5	6	2005	98,963.25	494.82	494.82	2,489.63
6	7	2005	97,910.10	1,547.97	494.82	2,984.45
7	8	2005	97,910.10	489.55	489.55	3,474.00
8	9	2005	97,910.10	489.55	489.55	3,963.55
9	10	2005	96,841.88	1,557.77	489.55	4,453.10
10	11	2005	96,841.88	484.21	484.21	4,937.31
11	12	2005	96,841.88	484.21	484.21	5,421.52
12	1	2006	95,757.56	1,568.53	484.21	484.21

6. Many of the calculators on the Web will let you solve this problem directly: entering what you know and getting what you want. My spreadsheet is set up to calculate the payment from the other variables. To solve this problem on my spreadsheet, you'll have to enter the number of monthly payments and the rate,

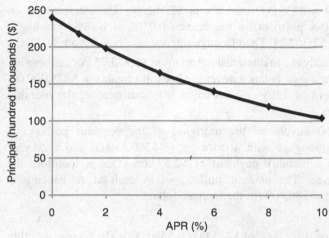

Figure 16.4

and then adjust the principal until the payment gets very close to $1,000. This problem illustrates the importance of getting as low an interest rate as you can to maximize your borrowing power for a fixed payment amount.

See Figure 16.4.

7. The easiest way to look at this is to say that you're paying the $250 up-front costs and therefore only putting $4,750 down on the car. The principal on your loan is therefore $31,800 − $4,750 = $27,050.

Pmt Nr	Mnth	Year	Balance ($)	Payment ($)	Interest ($)	Tot Int/Year ($)
1	5	2008	26,406.01	643.99	0.00	0.00
2	6	2008	25,916.06	643.99	154.04	154.04
3	7	2008	25,423.24	643.99	151.18	305.21
4	8	2008	24,927.56	643.99	148.30	453.51
5	9	2008	24,428.98	643.99	145.41	598.93
6	10	2008	23,927.49	643.99	142.50	741.43
7	11	2008	23,423.08	643.99	139.58	881.00
8	12	2008	22,915.72	643.99	136.63	1,017.64
9	1	2009	22,405.41	643.99	133.68	133.68
46	2	2012	1,276.80	643.99	11.14	25.95
47	3	2012	640.25	643.99	7.45	33.40
48	4	2012	0.00	643.99	3.73	37.13

16.4 CHAPTER 4

1. Three points on a $325,000 loan is 3%($325,000) = $9,750. The total up-front cost is $9,750 + $450 = $10,200. The monthly payment on a loan of

$325,000 + $10,200 = $335,200 at 5.8% is $2,362.96. The interest rate on $325,000 to match this payment, to the nearest 0.01%, is 6.18%, yielding a monthly payment of $2,362.28. The effective interest rate is therefore 6.18%.

2. The payments are exactly the first interest calculation: $1,673.75. Since these are interest-only payments, your balance never changes. It remains at $325,000.00. At the end of 3 years (the thirty-seventh payment), your new regular monthly payment is $2,577.66.

3. The balance after 60 months of the mortgage of the previous problem is $301,971.35. A 5% mortgage with a principal of $390,580.00 and a 20-year payment period has a monthly payment of $2,577.66. This is therefore the desired new mortgage. The amount pulled out in cash at refinancing is $390,580 − $301,971 = $88,609 (to the nearest dollar).

4.

(a) The first mortgage is for 80% of $425,000 = $340,000. The regular monthly payment is $1,825.19. After 7 years, the outstanding balance is $299,013.05.

(b) The 3% a year appreciation is a compounding situation with 3% annual compounding. After 7 years, therefore, the house is appraised at $425,000 $(1 + 0.03)^7 = $522,696.40.$ (You can do this calculation on a pocket calculator or a blank spreadsheet, or by treating it as a compound interest problem using any compound interest calculator.)

(c) Eighty percent of the new house value is 80%(522.696.40) = $418,157.10. The second mortgage is a 30 − 7 = 23-year mortgage.

(d) The second mortgage principal is the difference between 80% of the new value and the balance of the first loan, which is $418,157.10 − $299,013.05 = $119,144.05. A 23-year mortgage at 6.2% annual percentage rate (APR) has monthly payments of $811.20.

(e) The total monthly payment is $1,825.19 + $911.20 = $2,736.39.

16.5 CHAPTER 5

1.

(a) Using my spreadsheet, the worst case penalty occurs after monthly payment 21. It is $93.58.

(b) It would cost the payoff amount of $17,303.98 plus the monthly payment of $502.14 plus the penalty of $93.58 = $17,899.69.

(c) Before: $93.58/($502.14 + $17,303.98) = 0.53%.
After: $93.58/17,303.98 = 0.54%.

(d) After payment 21, the remaining balance is $17,303.98. One month's interest on this balance is $112.06, so 3 month's interest is $336.18. Compared with the rule of 78 penalty, this is 336.18/93.58 = 3.59, which is 259% larger.

2. The strategy here will be to consider all the different options and compare the savings account balance at a time when all of the options have the loan fully paid off, that is, at the end of 60 months.

First, consider the case where there is no prepayment penalty. If the loan is paid off after, say, the third payment, the savings account balance at that time is calculated by considering a $100,000, 7.00% APR account, which is compounded monthly and has $502.14 withdrawn monthly:

Pmt Nr	Balance ($)
0	100,000.00
1	100,018.19
2	100,162.86
3	100,245.00

At that time, the loan balance is $23,962.04. After paying off the loan, the savings account balance is $100,245.00 − $23,962.04 = $76,282.96. This balance then earns 7.00% interest for the remaining 57 months:

$$(\$76,282.96)(1 + 0.07/12)^{57} = \$106,270.06.$$

Showing the results for a few of these calculations:

Payoff month	Balance at the end of 60 months ($)
0	106,321.89
1	106,304.27
2	106,287.00
3	106,270.06
4	106,253.46
59	105,813.12
60	105,812.87

The last row of the table, payoff month 60, is identically the case when the loan is not prepaid.

The best case is to pay the loan off as quickly as possible—each month that you delay leaves you with a little less money at the end of the 5-year period.

If you must pay a prepayment penalty, then your only option from the above list is to never take the loan (payoff month = 0) in the first place or, having taken the loan, not to pay it off early. Since the former case is better for you, the benchmark of $106,321.89 is what you're trying to beat.

The following is generated in the same manner as the previous table. The only difference is that in each case, the loan balance is augmented by the prepayment penalty when the loan is paid off.

Payoff month	Balance at the end of 60 months ($)	
	Rule of 78	3% of Rem Bal
0	106,321.89	105,648.52
1	106,290.45	105,644.01
2	106,260.39	105,639.78
3	106,231.67	105,635.81
59	105,812.76	105,813.12
60	105,812.87	105,812.87

Regardless of whether you're dealing with the rule of 78 or the 3 months' interest on the remaining balance prepayment penalties, once you've signed the loan, you want to pay off the loan as quickly as you can get down to the bank.

3. In this case, I don't have to show any calculations. If I'm earning more than my loan is costing, I just make my loan payment every month. There is no point to paying off the loan. While I have the loan, I still have the remaining loan balance earning 8.0% while I'm paying 7.6% on the loan. Prepayment penalties never come into the picture because I have no intention of paying off the loan early.

This situation can be generalized: If you can earn a higher interest rate on borrowed money than you are paying on this borrowed money, borrow as much as you can and use it to start earning. Whether you pay off your loan month by month or all at the end of some period, you are making money every month.

16.6 CHAPTER 6

1.

(a) $150.00

(b) In the month of your purchase, your daily balance every day equals your average daily balance, which is $150.00. Your interest for the month is then ($150)(0.000333)(30) = $1.50. Your statement bill is for $150.00 if you pay before the due date, but since you don't, your new daily balance on the first day of the next billing period is $151.50. On the second day, your payment is credited and your daily balance drops to $51.50.

(c) Your daily balance will remain at $51.50 until the end of this billing period. Your average daily balance at that time will be

$$\text{ADB} = \frac{\$151.50 + 30(\$51.50)}{31} = \$54.73.$$

The interest on this average daily balance is ($54.73)(0.000333)(31) = $0.57, and your purchase balance is then $51.50 + $0.57 = $52.07.

2. I'm assuming that there are no other balances on your card and that you're not using your card for anything else.

Starting with the first month that you do this, from the first through the nineteenth of this month, your daily balance is 0. From the twentieth to the thirty-first of this month your daily balance is $250.00. Your average daily balance is

$$\frac{19(\$0.00)+12(\$250.00)}{31} = \$96.77,$$

and your interest is $96.77(0.0005)(31) = $1.50. Your bill for the first month is $250.00 + $1.50 = $251.50.

For the first 2 days of the second month, your daily balance is $251.50. On the third day, your payment for this amount reaches the bank and your daily balance drops to 0, where it remains until the twentieth of the month when it jumps to $250.00 and stays there for the rest of the month. Your average daily balance for the second month is then

$$ADB = \frac{2(\$251.50)+17(\$0.00)+12(\$250.00)}{31} = \$113.00.$$

Your interest for the second month is $113.00(0.0005)(31) = $1.75, and your bill for the second month is $250.00 + $1.75 = $251.75. This can be continued for as many months as you wish.

3. Starting on the first day of the month, you have a purchase daily balance (PDB) of $150.00 and a cash advance daily balance (CDB) of 0. On the twentieth of the month, the CDB jumps to $250.00. There are no other changes in the daily balances for the rest of the first month.

On the last day of the first month, the average daily balances are

$$\text{Purchases (PADB)} = \frac{31(\$150.00)}{31} = \$150.00$$

and

$$\text{Cash advance (CADB)} = \frac{19(\$0.00)+12(250.00)}{31} = \$96.77.$$

Your finance charges (interest) are the following:

Purchase interest = $150.00(0.0003)(31) = $1.40.

Cash advance interest = $96.77(0.0006)(31) = $1.80.

Your end of first month balances are the following:

Purchases: $150.00 + $1.40 = $151.40.

Cash advance: $250 + $1.80 = $251.80.

Total: $151.40 + $251.80 + $403.21. This is the amount on your statement.

For the first 2 days of the second month, PDB is equal to $151.40 and CDB is equal to $251.80. On the third day of the second month, your payment of $50 of your first month's statement, which is 50%($403.21) = $201.61, arrives. This amount is first applied to your PDB, reducing it to 0, and then the remainder of your payment, $206.61 − $151.40 = $55.21, is applied to your CDB, reducing it to $251.80 − $55.21 = $196.59. These daily balances remain the same until the end of the second month.

The average daily balances for the second month are

$$PADB = \frac{2(\$151.40) + 29(0)}{31} = \$9.77$$

and

$$CADB = \frac{2(\$251.80) + 29(\$196.59)}{31} = \$193.67.$$

The finance charges for the second month are

Purchase interest = $9.77(0.0003)(31) = $0.09 and

Cash advance interest = $193.67(0.0006)(31) = $3.60,

and for the end of the month balances are the following:

Purchases: $0.00 + $0.09 = $0.09.

Cash advance: $196.59 + $3.60 = $200.19.

Total: $200.38. This is the amount on your statement.

16.7 CHAPTER 7

1. $22,000

2. The present value of 24 monthly payments of $100 at different APRs is

APR (%)	PV ($)
0	2,400
2	2,355
4	2,311
6	2,268

If I can save or invest my money at greater than approximately 4%, it's a good deal for me. In this problem, practical considerations might outweigh numerical results. If I can only get a 2% APR on my savings, then I am "overpaying" about $55 on a $2,300 purchase. This is annoying but probably won't change my life. If this purchase was for, say, a recreational vehicle costing $230,000 and the monthly payment was $10,000, I would probably be more concerned about the present value.

16.8 CHAPTER 8

1. Generate this information using the spreadsheet Ch8LoanPV.xls. Set an interest rate; set the savings rate (PV Rate on the spreadsheet) to one-half that value; and tweak the principal until the Total PV is about $400,000:

Rate (%)	PV Rate (%)	Principal ($)	Total PV ($)
0	0	400,000	400,000
4	2	348,000	400,012
5	2.5	337,300	400,029
6	3	327,500	400,189
10	5	294,400	400,059

2.

Rate (%)	Principal ($)	Payment ($)	Ratio
0	400,000	2,222	180
2	372,000	2,394	155
4	348,000	2,574	135
6	327,500	2,764	118
8	309,600	2,959	105
10	294,400	3,164	93

From the given data, it's pretty clear that the ratio of principal to payment goes down as the loan APR goes up. In other words, even though all of the loans in the example have the same present value, you get a lot more loan for your dollar payment when the interest rate is low than when it is high.

3.

Nr Pmts	Loan rate (%)	Savings rate (%)	PV ($)	Monthly payment ($)	Figure of merit
120	6.00	3.50	392,950	3,886	153
120	7.00	3.50	410,958	4,064	167
120	8.00	3.50	429,431	4,246	182
240	6.00	3.50	432,359	2,508	108
360	6.00	3.50	467,309	2,098	98
240	7.00	3.50	467,885	2,714	127
240	8.00	3.50	504,783	2,928	148
360	7.00	3.50	518,558	2,329	121
360	8.00	3.50	571,920	2,568	147

Discussion: Using these data, the lowest figure of merit (FM) is 98. This is for the loan with the lowest payment of the group. The loan with the lowest PV is at the top of the table, with an FM of 153. The arguments made in this chapter for preferring loans with low present values are solid. On the other hand, multiplying together two numbers that we want to minimize and looking at the product are arbitrary procedures with no real merit. The loan with this FM of 98 has very low payments because there are many of them. It's still an expensive loan.

4. Eight years into the mortgage is the ninety-sixth payment. At that time, your balance is $195,825.35 (I'm using the ARM tab of Ch8LoanPV.xls). If a lender is offering you a 4.2% mortgage at that time, you can assume that savings interest rates have also dropped and are now at about 2.1%. Your two possible scenarios are the following:

(a)

			Tot PV:	$216,036.69		
Nr Pmts:	84	Pmt Nr	Balance ($)	Payment ($)	Interest ($)	PV ($)
Principal:	$195,825	0	195,825.00	0.00	0.00	0.00
Rate:	5.00%	1	193,873.16	2,767.77	815.94	2,762.94
		2	191,913.20	2,767.77	807.80	2,758.11
PV Rate:	2.10%	3	189,945.06	2,767.77	799.64	2,753.29
		4	187,968.73	2,767.77	791.44	2,748.48
Up Front:	$0	5	185,984.16	2,767.77	783.20	2,743.68
		6	183,991.32	2,767.77	774.93	2,738.89

(b)

			Tot PV:	$215,377.44		
Nr Pmts:	96	Pmt Nr	Balance ($)	Payment ($)	Interest ($)	PV ($)
Principal:	$195,825	0	195,825.00	0.00	0.00	0.00
Rate:	4.20%	1	194,105.16	2,405.23	685.39	2,401.02
		2	192,379.30	2,405.23	679.37	2,396.83
PV Rate:	2.10%	3	190,647.40	2,405.23	673.33	2,392.64
		4	188,909.44	2,405.23	667.27	2,388.46
Up Front:	$3,000	5	187,165.40	2,405.23	661.18	2,384.29
		6	185,415.25	2,405.23	655.08	2,380.13

The new loan is a slightly better deal (PV of $215,377 as compared with $216,036). The new payments are lower, but you're paying for an extra year.

5. Before going to the spreadsheet, you have to decide what the savings APR should be. It doesn't make sense to estimate it to be one-half of the loan APR when

you're looking at three different loan APRs. Just to cover our bets, I'll set it up for a few different savings APRs:

Lender	PV at Savings APR ($)			
	2.00%	2.50%	3.00%	3.50%
A	514,130	480,948	450,737	423,193
B	275,460	257,689	241,502	226,744
C	855,025	799,841	749,599	703,792

Regardless of the savings APR chosen, loan C has a higher present value than the sum of the APRs of loans A and B. The differences are extreme enough that up-front cost considerations won't change the conclusion—take loans A and B.

16.9 CHAPTER 9

1. For each of you:

   Single: $4,481.25 + 0.25($50,000 − $32,550) = $8,843.75.

   Married filing separately: same.

   For both of you, with a total taxable income of $100,000:

   Married filing jointly: $8,962.50 + 0.25($100,000 − $65,100) = $17,687.50, which is exactly twice $8,843.75. It would appear that there is no advantage to either situation.

2. Married filing jointly, the taxable income is $100,000 and, as above, the tax is $17,687.50.

   Single: $10,000 income is $802.50 + 0.1($10,000 − $8,025) = $1,000.00.

   $90,000 income is $16,056.25 + 0.28($90,000 − $78,850) = $19,178.25.

   Total tax: $1,000 + $19,178.25 = $20,178.25.

   Married filing separately:

   $10,000 income is $802.50 + 0.15($10,000 − $8,025) = $1,098.75.

   $90,000 income is $12,775 + 0.28($90,000 − $65,725) = $19,572.00.

   Total tax: $20,670.25.

3. I'll start with the loan information. I'm not interested in the payments right now, but in the interest totals for the year:

Year	6.0%	6.5%
	15 years	20 years
1	18,914	20,368
2	19,579	21,950

Continued

Year	6.0%	6.5%
	15 years	20 years
3	18,792	21,323
4	17,765	20,654
5	16,674	19,940
6	15,517	19,178
7	14,288	18,365
8	12,983	17,498
9	11,598	16,573
10	10,127	15,585
11	8,566	14,532
12	6,908	13,408
13	5,148	12,209
14	3,280	10,929
15	1,296	9,564
16	8,108	8,108
17		6,554
18		4,895
19		3,126
20		1,238

Also, the total present value of these loans is $427,681 and $470,522 for the 6.0% and 6.5% loans, respectively.

Next, I need to calculate your tax bill for the next 20 years with the interest deductions in the two scenarios. Then, I'll need the present value of your bills:

Year	Income ($)	Corr1 Inc1($)	Tax1($)	PV1($)	Corr2 Inc2($)	Tax2($)	PV2($)
1	50,000	31,086	3,860	3,860	29,632	3,642	3,642
2	51,500	31,921	3,986	3,870	29,550	3,630	3,524
3	53,045	34,253	4,335	4,087	31,722	3,956	3,729
4	54,636	36,871	4,728	4,327	33,982	4,295	3,930
5	56,275	39,601	5,138	4,565	36,335	4,648	4,130
6	57,964	42,447	5,565	4,800	38,786	5,015	4,326
7	59,703	45,415	6,010	5,033	41,338	5,398	4,521
8	61,494	48,511	6,474	5,264	43,996	5,797	4,713
9	63,339	51,741	6,959	5,493	46,766	6,212	4,904
10	65,239	55,112	7,464	5,721	49,654	6,646	5,093
11	67,196	58,630	7,992	5,947	52,664	7,097	5,281
12	69,212	62,304	8,543	6,172	55,804	7,568	5,467
13	71,288	66,140	9,223	6,468	59,079	8,059	5,653

Continued

Year	Income ($)	Corr1 Inc1($)	Tax1($)	PV1($)	Corr2 Inc2($)	Tax2($)	PV2($)
14	73,427	70,147	10,224	6,962	62,498	8,572	5,837
15	75,629	74,333	11,271	7,451	66,065	9,204	6,085
16	77,898	77,898	12,162	7,806	69,790	10,135	6,505
17	80,235	80,235	12,746	7,943	73,681	11,108	6,922
18	82,642	82,642	13,348	8,076	77,747	12,124	7,335
19	85,122	85,122	13,968	8,205	81,996	13,186	7,746
20	87,675	87,675	14,606	8,330	86,437	14,297	8,153

Tot PVs	$120,379	$107,498
Loan:	$427,681	$470,522
Grand total:	$548,060	$578,020

The total present value of all of your tax bills is lower for the 6.5% loan. However, it is not lower enough to outweigh the significantly lower present value of the 6.0% loan. Being able to deduct interest payments from your tax bill closes the gap between the present values of the two loans but not enough; the 6% loan is still the better deal.

4. I'll simplify a little and just do end-of-year accountings. In the first year, the $500,000 savings account grows by 5% to $525,000 and then loses $40,000 to withdrawals, leaving $485,000. This process repeats each year, with the cost of living growing every year while the balance falls every year until you run out of money in the fifteenth year:

Year	C.O.L. ($)	Balance ($)
1	40,000	485,000
2	41,200	468,050
3	42,436	449,017
4	43,709	427,758
5	45,020	404,126
6	46,371	377,961
7	47,762	349,097
8	49,195	317,357
9	50,671	282,554
10	52,191	244,491
11	53,757	202,959
12	55,369	157,737
13	57,030	108,594
14	58,741	55,282
15	60,504	−2,457

5. An amount of $100,000 taxable income is pretty close to the middle of the 25% tax bracket for married couples filing jointly. Since your interest is going to be in the several thousand dollar range, you will remain in this bracket. This means that your incremental tax rate is 25%; that is, 25% of all taxable savings or investment income goes to the IRS. Mathematically, you are keeping $100\% - 25\% = 75\% = 3/4 = 0.75$ of your taxable incremental income. This means that when you have a target number for what you want to keep, you have to earn $1/0.75 = 1.33$ times that amount.

If inflation is running at 2.5% then a tax-free investment must earn 4.5% in order for you to have 2% actual growth. A taxed investment or savings must earn $1.33(4.5\%) = 6.0\%$.

16.10   CHAPTER 10

1.

(a)

Age	q	Cost ($)	PV ($)
50	0.005648	5,648	5,538
51	0.006121	6,121	5,771

Since these are two separate policies, the cost of each is just the amount of the policy times the probability of death during that year (q). The present value of the first policy is just the cost 0.5 year earlier (on the man's fiftieth birthday). The present value of the second policy is just the cost 1.5 years earlier. The total present value is just the sum of the two individual present values, which is $11,309.

(b) The first year of this policy is identical to the 1-year policy above. However, since this is being written as a 2-year policy with the premium collected up front (on the man's fiftieth birthday), the second year of the policy costs a little less than it did above because we must account for the fact that the man might die during his fiftieth year:

Age	q	l	d	Cost ($)	PV ($)
50	0.005648	100,000	565	5,648	5,538
51	0.006121	99,435	609	6,087	5,739

In this table, we start with 100,000 people so that $l_{50} = 100,000$. The number of people that die that year is just $l_{50}q_{50} = 565$. The cost of a $1,000,000 policy is then $(565)\$1,000,000/100,000 = \$5,648$, which is identical to the cost in solution 1a. Continuing, $l_{51} = l_{50} - d_{50} = 99,435$ and then $d_{51} = l_{51}q_{51} = 609$, leading to a cost of $6,087. The present value of the second year's cost is $5,739.

2. In this situation, the person's age doesn't matter; we just have year 1 and year 2. Since the probability of living through year 1 is 2/3, the probability of dying during year 1 is $q_1 = 0.3333333$. Assuming the person survives to the second year, death is a certainty during this second year, so $q_2 = 1$. The 2-year Life Table is then

Year	q	l	d	Cost ($)	PV ($)
1	0.3333	100,000	33,333	33,333	32,686
2	1	6,667	6,667	66,667	62,858

As with any life policy that is in effect until the person dies, the sum of the costs must be the value of the policy. The present values are of course lower because of the time value of money—the insurance company is getting the money up front, before it has to pay out.

3. This problem can be solved in either of two ways. First, we use Table 10.6 exactly as we used the original Life Tables. Remember that in this situation, the first present value is one-fourth reflected one-fourth year back and so on:

Age	r	l	d	Cost ($)	PV ($)
48.5	0.002477	100,000	248	123.85	122.64
49.0	0.002581	99,752	257	128.72	124.99
49.5	0.002685	99,495	267	133.60	127.20

The total cost is the sum of the present values, which is $374.83

An alternate approach is to use the first line of the table above for the half year policy from 48.5 to 49.0 years back and then to use the original Life Table for a single-year term policy bought at age 48.5 for the year 49.0–50.0, noting that we are not starting with 100,000 people but with 99,752. In this case, the present value of this latter policy is the cost reflected back one full year:

Age	q	l	d	Cost	PV
49	0.005206	99,752	519	$259.67	$249.69

The total of the present values is $249.69 + $122.64 = $372.33.

The slight discrepancy in the two answers is analogous to the difference between APR and effective annual percentage rate (EAPR).

4. In the following table, for each group of 5 years, the first l value is the number of people alive at the start of that year. The d values for the 5-year group are the number of people who die during that 5 years. The sum of these five d values, divided by the first l value, is the probability of dying during those 5 years, assuming you reached the first of those 5 years alive:

Age	q	l	d	Sums	q Abridged
25–26	0.000506	98,710	50	269	0.002723
26–27	0.000522	98,661	51		
27–28	0.000541	98,609	53		
28–29	0.000565	98,556	56		
29–30	0.000593	98,500	58		
30–31	0.000627	98,442	62	353	0.003588
31–32	0.000667	98,380	66		
32–33	0.000712	98,314	70		
33–34	0.000764	98,244	75		
34–35	0.000825	98,169	81		
35–36	0.000892	98,088	88	533	0.005434
36–37	0.000971	98,001	95		
37–38	0.001071	97,906	105		
38–39	0.001190	97,801	116		
39–40	0.001321	97,684	129		
40–41	0.001453	97,555	142	846	0.008673
41–42	0.001586	97,414	154		
42–43	0.001727	97,259	168		
43–44	0.001883	97,091	183		
44–45	0.002055	96,908	199		

Extracting the information for the abridged table:

Age	q Abridged	l	d
25–30	0.002723	98,710	269
30–35	0.003588	98,442	353
35–40	0.005434	98,088	533
40–45	0.008673	97,555	846

5.

(a) The monthly payment is $5,311.76. At the beginning of each year, the balances are

Year	Balance ($)
1	250,000.00
2	209,432.97
3	164,618.04
4	115,110.41
5	60,418.67

(b)

Age	q	l	d	Policy ($)	Cost ($)	PV ($)
35	0.001653	100,000	165	250,000	413.16	403.20
36	0.001770	99,835	177	209,433	370.03	343.92
37	0.001911	99,658	190	164,618	313.55	277.54
38	0.002075	99,468	206	115,110	237.55	200.26
39	0.002254	99,261	224	60,419	135.17	108.53
				Tot	$1,333.46	
				Price	$1,800.17	

This table starts with the age, then the q values, for men age 35–39. l starts at $100,000$ ($l_{35} = 100,000$) and then $d_{35} = q_{35}l_{35} = 165$; $l_{36} = d_{35} - l_{35} = 99,835$; $d_{36} = q_{36}l_{36}$; and so on.

Since I started with 100,000 people, the cost of a $100,000 for any age ($i$) is just $d_i(\$100,000)/100,000 = d_i$. To get the cost of any other size policy, say, a $250,000 policy, just scale the numbers: cost for $250,000 policy at age 35 = 165($250,000)/100,000 = $413.16.

The present values are the values at the start of the policy, so the age 35 PV is the age 35 cost half a year earlier, which is $413.16/(1.05)^{0.5}$ and so on.

The total cost of the policy is the sum of all the present values, which is $1,333.46, and the insurance company price will be 35% large, which is (1.35)($1,333.46) = $1,800.17.

(c) Following the procedures of earlier chapters, folding this cost into the loan raises the payments to $5,350.01, which is equivalent to an effective interest rate on the original loan of 10.31%.

16.11 CHAPTER 11

1. From Table 10.1, the 2004 U.S. Life Table for all men, your life expectancy is 10.7 years. From the Life tab of the Ch11fixedannuities.xls spreadsheet, your annuity will pay about $3,950 each month. Your exclusion ratio, up until (and if) you pass your expected date of demise, is 0.690, which means that 31% of your annuity income is taxable. This amounts to 0.31(12)(3,950) = $14,700.

The $150,000 in your savings bank pays 0.05($150,000) = $7,500 in income.

Your total income for tax purposes is therefore $35,000 + $14,700 + $7,500 = $57,200. Your taxable income is $47,200. From Figure 9.1, Schedule X (single filing status), your federal tax is $4,481.25 + 0.25($57,200 − $32,550) = $10,640.

Your after-tax income is therefore $35,000 + $47,400 + $7,500 − $10,640 = $79,260 a year or $6,600 a month.

If you live longer than your expected death date, the entire annuity income becomes taxable: 12(3,950) = $47,400. Your total income, now all taxable, is $35,000 + $47,400 + $7,500 = $89,900 and your taxable income is $79,900.

From the tax table, your tax is $16,056.25 + 0.28($79,900 − $78,850) = $16,350. This leaves an after-tax income of $89,900 − $16,350 = $73,550 or $6,130 a month.

Note that while I used the Life tab of the spreadsheet to calculate the annuity properties, I didn't use its tax calculations. The spreadsheet can only calculate the incremental tax in a given bracket due to the annuity. It doesn't know your whole story.

2. From Table 11.4, for a 75-year-old woman and a 70-year-old man, the expected first death is in 8.5 years, the expected second death is in 16.9 years, or 8.4 years after the first death. Going to the Life tab on Ch11fixedannuities.xls, first note that the value put in for Age doesn't matter here—it's just used to predict the expected age at death and doesn't affect the premium calculation. Also, we're not interested in the tax calculations:

Age	0	Nominal number of monthly payments:	102
Multiplier	8.5		
		Calculated principal:	$200,367.28
Payment	$2,500	Exclusion ratio:	0.786
Rate	6.00%	Expected age at death:	8.5
Tax rate	0%	Tax fraction before expected death age:	0.214
		Monthly taxes before expected death age:	$0.00
		Monthly taxes after expected death age:	$0.00

The premium is approximately $200,400 for the first part of this package. We need a second annuity 8.5 years from now, for $2,000 a month with an expected term of 8.4 years. Using the same spreadsheet, the premium on this annuity is approximately $158,800. However, since we are paying the total premium today, this part has 8.5 years to grow to this value, at 6% APR. The present value is therefore $158,800/(1.06)^6, approximately $111,900.

Adding the two amounts that are to be spent today, $200,400 + $111,900 = $312,300.

3. From the Life Tables, an 82-year-old man has a life expectancy of about another 7 years. For all reasonable calculations, 18 years is the longest life expectancy to worry about. Using the Save tab of the Ch3Amortization.xls spreadsheet, think of having your daughter advance you regular monthly payments as if you were a savings bank. The interest rate should be the rate she's giving up by taking the money from her savings or investments. If you take $500 a month at 6% interest,

then after 7 years, the balance is about $52,000. After 18 years, the balance is about $177,000. At the time of your death, your daughter can repay herself from your life insurance policy and still have an inheritance. At 10% interest, after 18 years, the balance is about $266,000, slightly more than the value of your life insurance policy. However, if the interest rate when you start out is down around 5% and slowly drifts up to 10% over the 18-year period, then it's unlikely that the balance will exceed $250,000.

4. The first thing we have to do here is to estimate how many payments you will receive. Unless you know of some particular health or hereditary issues, the best you can do is to go to the Life Tables. Using the woman tables, we have

Age	e	Death
62	22.4	84.4
66	19.2	85.2
70	16.2	86.2

Since the Life Table expresses ages in tenths of a year, I assumed that social security payments come 10 times a year rather than monthly. This doesn't affect the results here because we're only interested in comparing the numbers, not the actual numbers. I could elect to receive $0.75 per payment starting when I'm 62 and expect to receive 224 of such payments, or to receive $1.00 starting when I'm 66 and expect to receive 192 payments, or $1.32 when I'm 70 and expect to receive 162 payments.

In this table, I'm calculating the present value of every payment on my sixty-second birthday, using an APR of 4.5%, compounded 10 times a year:

Age	PV of payment starting at age ($)		
	62	66	70
62	0.750	0.000	0.000
62.1	0.747	0.000	0.000
62.2	0.743	0.000	0.000
62.3	0.740	0.000	0.000
62.4	0.737	0.000	0.000
65.7	0.635	0.000	0.000
65.8	0.632	0.000	0.000
65.9	0.630	0.000	0.000
66	0.627	0.836	0.000
66.1	0.624	0.832	0.000
66.2	0.621	0.828	0.000

Continued

Age	PV of payment starting at age ($)		
	62	66	70
69.7	0.531	0.708	0.000
69.8	0.528	0.705	0.000
69.9	0.526	0.701	0.000
70	0.524	0.698	0.922
70.1	0.521	0.695	0.918
70.2	0.519	0.692	0.913
84.3	0.276	0.367	0.485
84.4	0.274	0.366	0.483
84.5		0.364	0.481
84.6		0.363	0.479
84.7		0.361	0.476
84.8		0.359	0.474
84.9		0.358	0.472
85.0		0.356	0.470
85.1		0.354	0.468
85.2		0.353	0.466
85.3			0.464
85.4			0.462
85.5			0.460
85.6			0.457
85.7			0.455
85.8			0.453
85.9			0.451
86.0			0.449
86.1			0.447
86.2			0.445

In the column for the payments shown starting at age 66, at age 66 is the present value of $1.00 calculated at age 62. Similarly, in the column for the payments starting at age 70, at age 70 is the payment of the present value of $1.32 calculated at age 62. The sums of each of the three columns are $106.45, $108.11, and $106.78.

According to this calculation, it hardly matters at all which choice you make. However, I just guessed at an APR and I ignored inflation.

A more interesting calculation is to assume a rate of inflation, an interest rate that stays about 2% above inflation, and a social security payment cost of living adjustment that tracks with inflation. Without showing the table here, for an inflation rate of 2.5% and an interest rate of 4.5%, the present values are (rounded a bit) $136 for starting at age 62, $148 for starting at age 66, and $157 for starting at age 70.

If we assume that the social security cost of living adjustment won't really keep up with inflation but will only account for about half the inflation rate, then the numbers become $120, $126, and $129, respectively. From these numbers you would conclude that you're better off waiting until age 70 to start collecting your social security, but this conclusion isn't compelling.

Before making a decision based on the above numbers, I recommend that you get the latest Life Tables that are appropriate for you and factor in any relevant family history and personal health factors that you are aware of.

16.12 CHAPTER 12

1. The present value of $1 10 years from now, at 7% interest (compounded annually), is about $0.50. In other words, for every dollar you loan to your customer today, he or she will owe $2 in 10 years. If you loan him or her half the appraised value of his or her house, and housing prices drop in half over the next 10 years, there will still be enough money to pay back the loan by selling the house. This means that the loan can be for $225,000. Since you are charging an up-front fee of $25,000, the customer can actually receive a check for $200,000.

What if the customer lives longer than age 87? This is where some historical perspective is useful. Home values have historically climbed. If home values climb by an average of 2% for 20 years and then suddenly plummet to two-third of their value, the home value returns to about today's price. If the customer lives to age 97, his or her outstanding loan balance will be about $450,000 on a $225,000 loan, and there is still enough value in the house to cover the loan.

A real business plan would need a deeper look at this, with real statistics covering many clients and various scenarios of housing values, but for a first pass, these numbers give a feeling of how well the loan is secured by the home.

Now, what about your company's cash flow? You had to borrow $225,000 today at 5% APR. You'd also like to make some profits over the life of the loan.

Year	Profit ($)	Balance ($)
1	8,000	233,000
2	8,000	252,650
3	8,000	273,283
4	8,000	294,947
5	8,000	317,694
6	8,000	341,579
7	8,000	366,658
8	8,000	392,990
9	8,000	420,640
10	8,000	449,672
11	8,000	480,156

Continued

Year	Profit ($)	Balance ($)
12	8,000	512,163
13	8,000	545,772
14	8,000	581,060
15	8,000	618,113
16	8,000	657,019
17	8,000	697,870
18	8,000	740,763
19	8,000	785,801
20	8,000	833,091

This table is very approximate but shows the business plan. On the first year of the loan, you borrow an additional $8,000, which you contribute to company profits (perhaps given out as dividends?). You are paying 5% on this money. At the end of 10 years, you owe about $450,000—the due amount of the loan. At the end of 20 years, you owe only about $786,000, much less than the due amount of the loan. It looks like you have about a $114,000 buffer here to further protect your company against unanticipated dips in home values.

2. Over the course of these years, it's pretty certain that interest rates will vary. This means that you might have to pay more on your loan waiting while waiting for your customer's loan to be repaid. You handle this by writing the original loan as an adjustable rate loan; when the rate you have to pay goes up, so does the rate your customer has to pay (and vice versa).

16.13 CHAPTER 13

1. The first thing to do is to normalize the prices of all the stocks, that is, divide each stock price by that same stock's price at the beginning of the year. In the table below, I created three idealized variations of ±5%, ±10%, and ±20% for the year:

Month	A	B	C	$1 buys		
				A	B	C
1	1.05	1.1	1.2	0.952	0.909	0.833
2	0.95	0.9	0.8	1.053	1.111	1.250
3	1.05	1.1	1.2	0.952	0.909	0.833
4	0.95	0.9	0.8	1.053	1.111	1.250
5	1.05	1.1	1.2	0.952	0.909	0.833
6	0.95	0.9	0.8	1.053	1.111	1.250

Continued

Month	A	B	C	$1 buys		
				A	B	C
7	1.05	1.1	1.2	0.952	0.909	0.833
8	0.95	0.9	0.8	1.053	1.111	1.250
9	1.05	1.1	1.2	0.952	0.909	0.833
10	0.95	0.9	0.8	1.053	1.111	1.250
11	1.05	1.1	1.2	0.952	0.909	0.833
12	0.95	0.9	0.8	1.053	1.111	1.250
	Average for the year:			1.003	1.010	1.042

Then the table shows what a $1 purchase will buy in terms of shares of each stock for each month. At the bottom of the table is the average amount of stock I got over the year for each dollar spent. Stock C is clearly the winner here. Also, since the average of a list of different numbers cannot be as large as the largest number in the list, spending all of my money on stock C is both better than spending it on any other stock and better than spending it on partial investments in each stock. Keep in mind that this idealized example ignores volatility and the advantages of diversification.

2. The correlation coefficient is 0.770. These stocks are moderately well correlated. The slope of the best-fit line is positive, indicating that these stocks tend to move together from day to day—perhaps, they're stocks for the same or related industries. The normalized standard deviation for stock #2 is almost twice that of stock #1, indicating that stock #2 is more volatile than stock #1. Both numbers are relatively small, however, indicating a relatively low level of volatility.

When the stock #1 numbers are replaced with the numbers 1–25, all information about stock #1 is lost. What we have left is a list of stock #2 prices versus the (relative) date when the number was collected. The slope of the best-fit line is a small positive number, indicating that this stock's price has, on the average, climbed over the time period represented. The correlation coefficient is very high, indicating that the stock's growth over this time period grew almost as a straight line graph. All other information generated is spurious.

3. Since you were given the options, your cost is 0. There is no reason to exercise these options unless the stock price is higher than the strike price. Figure 16.5 shows that your profit will be $1 per share for every $1 the stock price is higher than the strike price, and 0 otherwise. Exactly when you exercise the option is of course a guessing game. You'd like to wait until the stock price is at its highest, but you don't know when that will be, and at some time, the options will expire.

4. Figure 16.6 shows your strategy for this situation. You can only make a profit when the stock price is above $14, unlike in problem 1. However, if you're

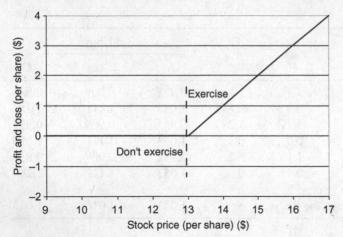

Figure 16.5

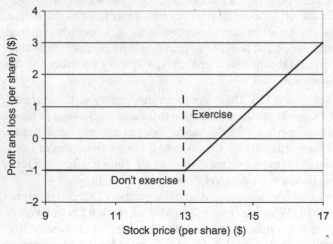

Figure 16.6

convinced that the stock price will never exceed $14 before your options expire, then you'll still exercise the option when the stock price is above $13. In the stock price region between $13 and $14, you will lose money, but you'll lose less than $1 per share, so you're at least minimizing your losses. If the stock price never gets above $13, then you freeze your losses at $1 per share by letting the option expire.

5. If the stock price never gets above $13, then nobody exercises their calls. You get to keep your $1 per share. If the stock price gets above $18, then you can expect your buyer to exercise. You exercise your calls, buy the stock at $13 a share, and sell it to your buyer at $18 a share. You earn $5 a share ($18 − $13) plus the $1 you already collected, making it $6.

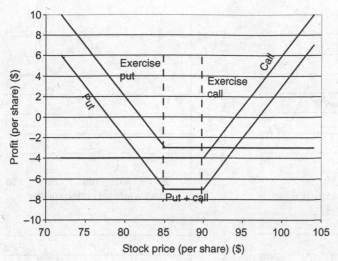

Figure 16.7

For the stock price between $13 and $18, different things could happen. Your buyer won't exercise his or her shares. If the stock price reaches, say, $15 a few hours before everything expires, you could take a chance: Exercise your options and sell the stock. You're betting that the stock won't cross $18 in the short time remaining. You earn the difference between the stock price and $13 plus the $1 you already have. If you bet wrong and the stock price shoots up past $18, you have to buy back the stock and deliver it to your call buyer. If, for example, the stock suddenly shoots up to $20, you have to buy back the stock that you sold for $15. Your loss is $5 a share minus the $1 you already collected, or $4 a share. If you're not a gambler, you just sit back and wait. When both sets of options expire, you still have the $1 you've already collected.

6. Figure 16.7 shows what you'd do with the call alone, what you'd do with the put alone, and with the combination of these two, which is simply a sum of the call and put graphs. For a stock price above $90, you would exercise the call and ignore the put. For a stock price below $85, you would exercise the put and ignore the call. You only make money if the stock price falls below $82 or climbs above $94 dollars. With this strategy, you are hoping for high volatility in the stock price but you don't really care which way the stock moves. If the stock stays between $82 and $94, you lose money, but the amount you lose can never be more than the sum of the put and call costs.

16.14　CHAPTER 14

The problems for Chapter 14 are just for fun. There's nothing to learn about games of chance other than that you can't beat the house. Games like poker combine chance, actual skill at playing hands, and psychological effects such as bluffing; they're very hard to quantify.

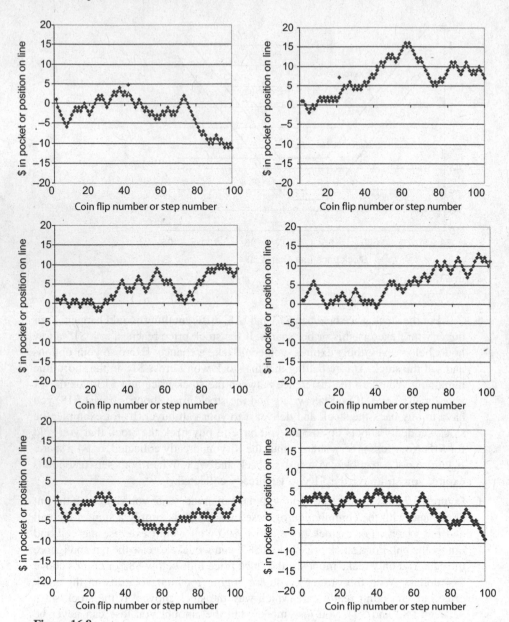

Figure 16.8

1. Figure 16.8 shows six runs of this spreadsheet "program" starting at +1 (or with $1 to play). If you were actually playing for money, as soon as you crossed below 0, you'd be wiped out. With a random walk, you just keep bouncing around.

   If you want an idea of what these graphs would be like starting at +5, or +10, or any number whatsoever, you don't have to recreate sample runs. All you

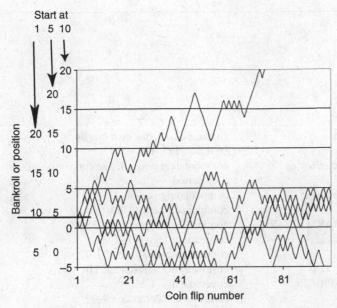

Figure 16.9

have to do is slide the vertical axis up or down until you have what you want. Figure 16.9 shows these results. There are vertical axes for starting out at +1, +5, and +10. Note that in all cases, it's not so unlikely that you get wiped out.

2.

(a) The state needs to sell 500 million tickets, so the probability of winning is 1/500 million. The probability of losing is 4,999,999/500,000,000, which is extremely close to one:

$$E = \$400,000 \left(\frac{1}{500,000,000} \right) - \$1 \left(\frac{499,999,999}{500,000,000} \right) \approx \$0.80 - \$1.00 = -\$0.20.$$

Another way of looking at this is that if you bought all 500 million lottery tickets, you would win the lottery and have lost $100 million (the state's profit). Each ticket would therefore have lost $100million/$500 million = 20 cents.

(b) About 1 second in 16 years.

Index